The rapid evolution of digital technology is changing how people see the world as they transform it. Geographic information system (GIS) professionals apply the tools of their trade in creative ways to solve problems, overcome challenges, improve productivity, and protect lives and property. Excellent examples of GIS in action across the globe are published in this 30th volume of the *Esri Map Book*. Geographic data is big data, and these examples show how our users analyze, visualize, and apply geographic data to create better understanding and a better future.

These maps demonstrate the integrative power of GIS to process vast, often diverse, amounts of data through web services, the cloud, and mobile devices. Whether applied to a health crisis like the Ebola outbreak, rising sea levels posed by climate change, or governments and businesses looking to be more efficient, the GIS platform is front and center. The app revolution helps professionals collaborate, support their customers, and increase productivity as they apply geographic information to design and decision-making processes. Nonprofessionals, too, are enjoying the benefits of increasingly user-friendly GIS, anytime and anywhere.

I am pleased that the *Esri Map Book* again includes the work of university students taking on such tasks as analyzing geologic features and mapping urban green space. The future of GIS is in good hands. I thank other users and organizations, many of them regular *Esri Map Book* contributors, for sharing their work to show that, today, geography and spatial thinking are more important than ever.

Warm regards,

Jack Dangermond

TABLE OF CONTENTS

2

MBR Global Consumer Styles

Michael Bauer Research GmbH

Nuremberg, Bavaria, Germany
By Michael Bauer Research GmbH

Contact
Sabine Petrasek
sabine.petrasek@mb-research.de

Software
ArcGIS for Desktop

Data Source
Michael Bauer Research GmbH

Michael Bauer Research GmbH (MBR) provides regional market data, such as purchasing power, sociodemographics, and consumer spending by product groups or retail data. MBR also provides the corresponding digital boundaries for administrative areas, ZIP/postal codes, or microgeographical level.

These maps show a new, worldwide segmentation approach based on a global survey with 10,000 respondents. The ten comparable global consumer styles are as follows:

  A. High-Earning Urban Professionals
  B. Comfortably Off Empty Nesters
  C. Modern and Pragmatic Over-50s
  D. Well-Informed Modern Consumers
  E. Affluent, Highly Educated Urban Families
  F. Security-Oriented Seniors
  G. Orientation-Seeking Lower- and Middle-Class Consumers
  H. Younger Lower- and Middle-Class Consumers
  I. Modern Younger Families
  J. Low-Income Younger Consumers

As the maps show, the "Security-Oriented Seniors" tend to live in rural areas, and "Affluent, Highly Educated Urban Families" are predominantly in economic centers and university cities.

Courtesy of Michael Bauer Research GmbH.

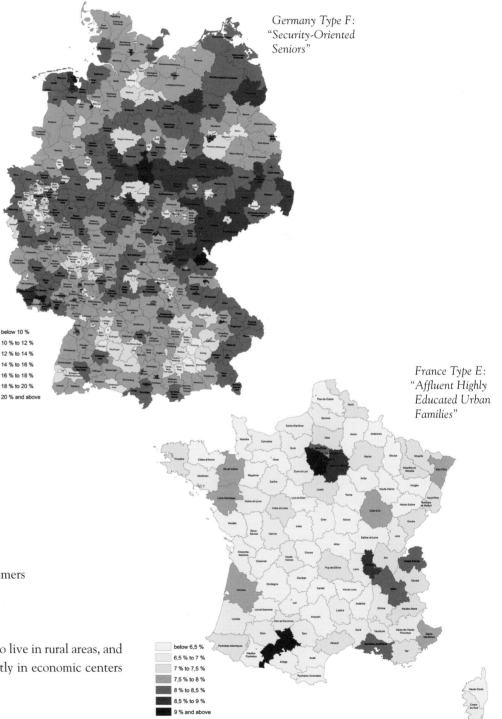

Germany Type F: "Security-Oriented Seniors"

France Type E: "Affluent Highly Educated Urban Families"

Where the Best Taxi Tippers Are in New York

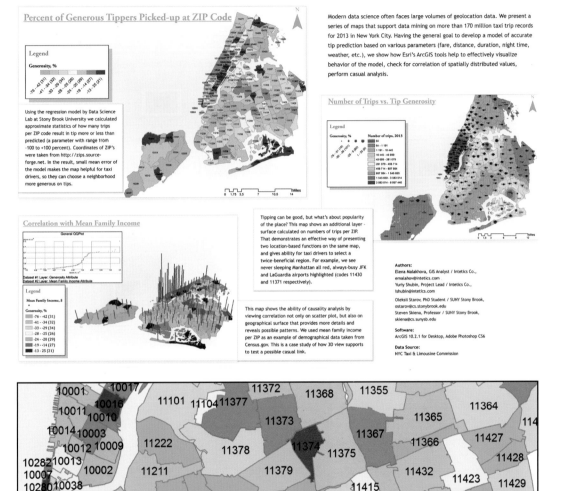

Modern data science often faces large volumes of geolocation data. We present a series of maps that support data mining on more than 170 million taxi trip records for 2013 in New York City. Having the general goal to develop a model of accurate tip prediction based on various parameters (fare, distance, duration, night time, weather, etc.), we show how Esri's ArcGIS tools help to effectively visualize behavior of the model, check for correlation of spatially distributed values, perform casual analysis.

Percent of Generous Tippers Picked-up at ZIP Code

Using the regression model by Data Science Lab at Stony Brook University we calculated approximate statistics of how many trips per ZIP code result in tip more or less than predicted (a parameter with range from -100 to +100 percent). Coordinates of ZIP's were taken from http://zips.source-forge.net. In the result, small mean error of the model makes the map helpful for taxi drivers, so they can choose a neighborhood more generous on tips.

Number of Trips vs. Tip Generosity

Tipping can be good, but what's about popularity of the place? This map shows an additional layer - surface calculated on numbers of trips per ZIP. That demonstrates an effective way of presenting two location-based functions on the same map, and gives ability for taxi drivers to select a twice-beneficial region. For example, we see never sleeping Manhattan all red, always-busy JFK and LaGuardia airports highlighted (codes 11430 and 11371 respectively).

Correlation with Mean Family Income

This map shows the ability of causality analysis by viewing correlation not only on scatter plot, but also on geographical surface that provides more details and reveals possible patterns. We used mean family income per ZIP as an example of demographical data taken from Census.gov. This is a case study of how 3D view supports to test a possible casual link.

Authors:
Elena Malakhova, GIS Analyst / Intetics Co., emalahov@intetics.com
Yuriy Shubin, Project Lead / Intetics Co., lshubin@intetics.com
Oleksii Starov, PhD Student / SUNY Stony Brook, ostarov@cs.stonybrook.edu
Steven Skiena, Professor / SUNY Stony Brook, skiena@cs.sunysb.edu

Software:
ArcGIS 10.2.1 for Desktop, Adobe Photoshop CS6

Data Source:
NYC Taxi & Limousine Commission

Intetics Ukraine and State University of New York at Stony Brook

Kharkiv, Ukraine and Stony Brook, New York, USA

By Elena Malakhova, Oleksii Starov, Yuriy Shubin, and Steven Skiena

Contact
Elena Malakhova
elenamalakhova88@gmail.com

Software
ArcGIS 10.2.1 for Desktop, Adobe Photoshop CS6

Data Source
New York City Taxi & Limousine Commission

Finding neighborhoods with generous taxi tippers is useful for any taxi company. Intetics, a global sourcing company, analyzed data about 170 million taxi rides in New York City in 2013 to determine the location of generous tippers. This map shows, by ZIP Code, the percentage of people taking a cab who pay tips that are more than a baseline amount (derived from a regression model that takes into account fare, distance, duration, time of day, and so on). The values range from -100 percent (all tip less) to +100 percent (all tip more). Intetics combined previous statistics with how many rides were started in a ZIP Code. This map shows that airports are popular for taxi pickup, but tips may not be that generous. Manhattan, in general, is busy and generous, especially some neighborhoods on the East Side. The map shows how GIS technology supports data science by displaying data, checking for correlation of spatially distributed values, and facilitating casual analysis.

Courtesy of Intetics Ukraine and State University of New York at Stony Brook.

Business License Locations within Murray City

Murray City

Murray City, Utah, USA
By Steve Kollman

Contact
Steve Kollman
skollman@murray.utah.gov

Software
ArcGIS 10.2.1 for Desktop

Data Sources
Murray City, Utah Automated Geographic Reference Center, Salt Lake County GIS

Murray City's economic development professionals provide real-time GIS analysis to select the best locations for businesses. Before GIS, finding business locations included several manual steps. Business license applications and business location inquiries had to be reviewed by city divisions that research zoning restrictions, engineering infrastructure improvements, code enforcement regulations, vehicle/pedestrian traffic impacts, and distances to similar businesses. This process could take weeks. Today, GIS answers complex questions pertaining to business license and location requests, often without delay at the time of the request. GIS also offers other beneficial information such as alternative locations, traffic studies, proximity to other commercial areas, public transportation facilities, and freeway access. GIS has greatly improved the efficiency of Murray City's business license request process and provides tools to attract new commercial opportunities.

Courtesy of Steve Kollman, Murray City, Utah.

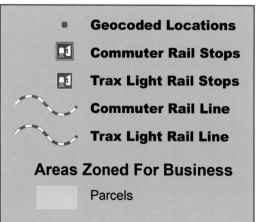

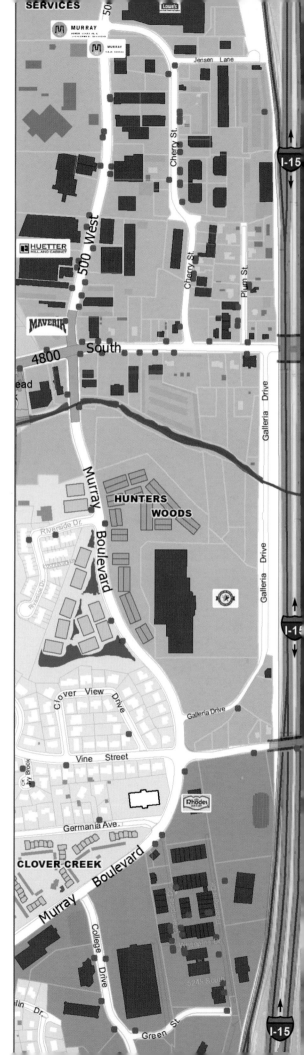

Swiss National Map 1:25,000

Federal Office of Topography (swisstopo)

Wabern, Bern, Switzerland
By Federal Office of Topography (swisstopo)

Contact
Urs Isenegger
Urs.isenegger@swisstopo.ch

Software
ArcGIS Desktop 9.3.1

Data Source
Federal Office of Topography (swisstopo)

The Federal Office of Topography (swisstopo) is beginning a comprehensive upgrade of the largest official map document for Switzerland—the 247-sheet, 1:25,000-scale national map. This is the most accurate and detailed topographic map of Switzerland for hikers, alpinists, vacationers, and adventurers. To produce the print and digital products, swisstopo is using data from the topographical landscape and the elevation model, automating cartographic production processes, and providing data in a flexible digital cartographic model. Swisstopo will replace the former national map with the updated map sheets as they are completed. The complete upgrade is scheduled to be completed in 2019. Automation will allow quicker updates to the national map in the future.

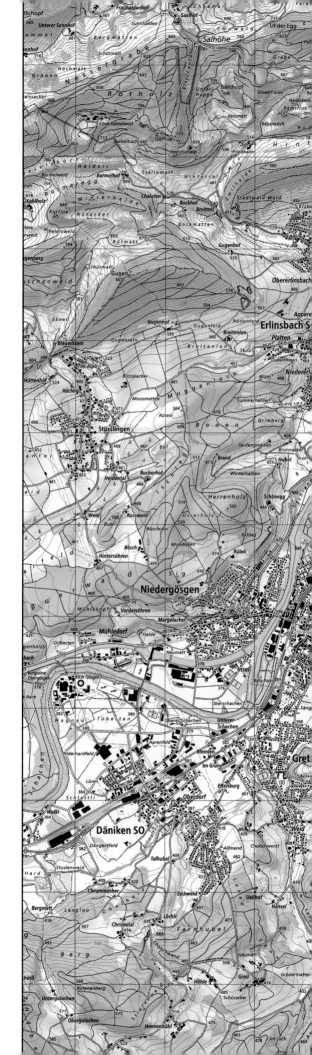

Swiss National Map 1:1 Million

swisstopo

Wabern, Bern, Switzerland
By Federal Office of Topography (swisstopo)

Contact
Urs Isenegger
Urs.isenegger@swisstopo.ch

Software
ArcGIS Desktop 9.3.1

Data Source
Federal Office of Topography (swisstopo)

The 1:1 million-scale national map of Switzerland is the first visible step in a new era: the production of topographic maps with the aid of database-supported GIS cartography. With this new map, the advantages of database-supported geodata have been successfully combined with the traditionally high cartographic and design requirements in Switzerland. The existing digital cartographic model "DKM1M" was produced using an ArcGIS-based application that was developed for swisstopo and can be used for developing a variety of products and services. The georeferenced pixel map and the full-color paper map printed form the basis. The method used to produce the 1:1 million-scale map is also to be used for other scales.

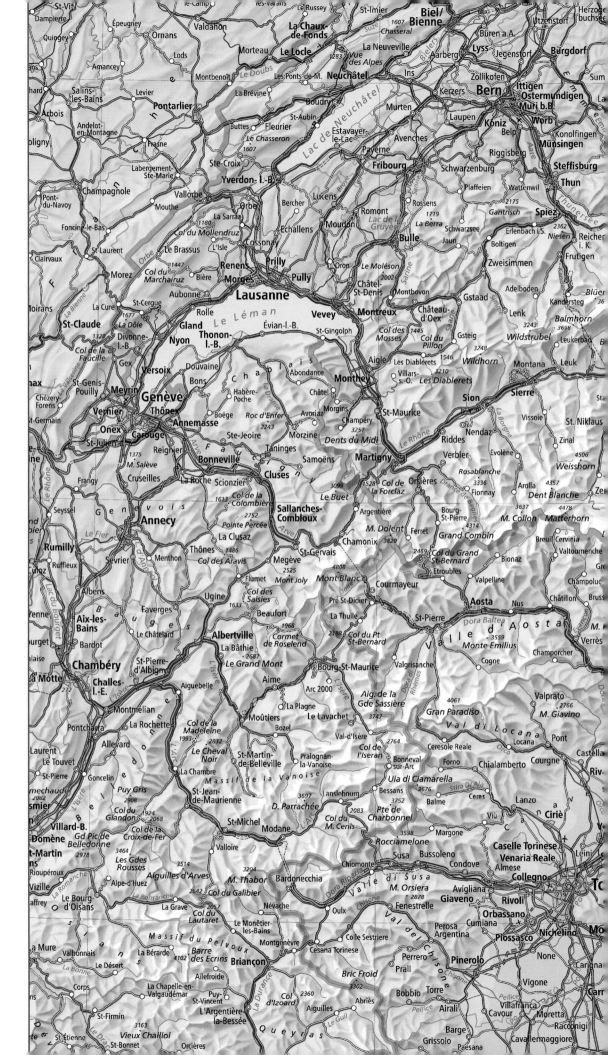

CARTOGRAPHY

Digital Tectonic Map of the World

Geologic Data Systems

Denver, Colorado, USA
By Geologic Data Systems

Contact
Katherine Connors
kconnors@gdata.com

Software
ArcGIS 10.3 for Desktop

Data Source
Geologic Data Systems

Geologic Data Systems' *Digital Tectonic Map of the World* is a new geologic and tectonic interpretation created specifically for use in the GIS environment. The design of the map is multilayered so no single, physical representation can display all of the information contained in the digital product. The intent is to provide users with a multitier digital dataset that allows them to display global tectonic settings most pertinent to their specific issue. This map shows tectonic elements, including Precambrian shields, fold belts, igneous terranes, and sedimentary basins. The map also shows ocean structural features, seafloor age, spreading centers, and other significant tectonic features.

Courtesy of Geologic Data Systems.

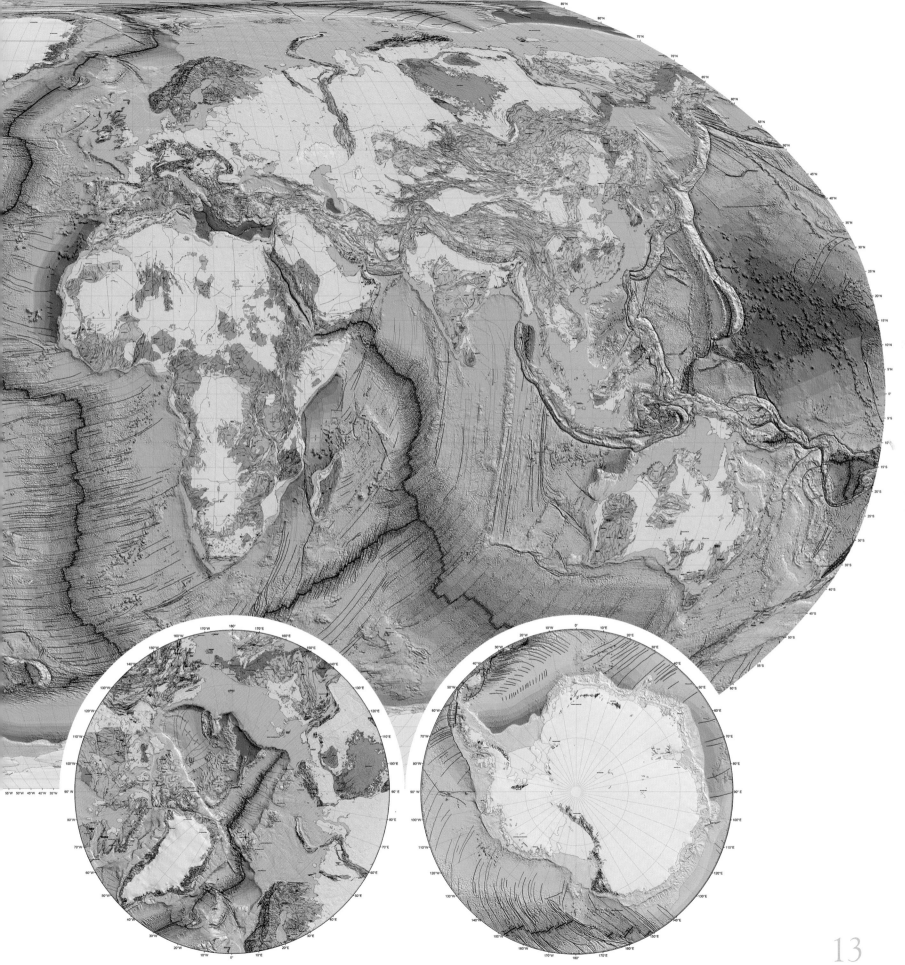

13

Tectonic Examples from the Greater Arctic Region

Geologic Data Systems

Denver, Colorado, USA
By Geologic Data Systems (GDS)

Contact
Katherine Connors
kconnors@gdata.com

Software
ArcGIS 10.3 for Desktop

Data Source
Geologic Data Systems

The Arctic Region panel of the GDS *Digital Tectonic Map of the World* includes a variety of possible data displays using subset layers from the complete dataset. The top row here shows three examples of primary data used in the interpretation. From top to bottom, these are Landsat imagery, elevation data, and gravity. The bottom row shows three possible combinations of the interpreted tectonic map data, including the following (from left to right):

- A map showing tectonic basins with sediment thickness isopachs, spreading ridges, ocean structure, and the ocean elevation data colored by depth.
- A representation of the tectonic geology (rock exposures that define tectonic elements) along with seafloor age, onshore and offshore structure and the passive margin region shown by the pale orange hatch pattern.
- A map that displays the onshore hillshade draped with the Landsat imagery, a deep red stipple pattern showing large igneous provinces, ocean floor colored using a gradational scheme from warm to cool colors with increasing age, the passive margin area shaded in yellow, and the type of passive margin (undifferentiated or with volcanic seaward dipping reflectors) indicated by the bounding line type.

Courtesy of Geologic Data Systems.

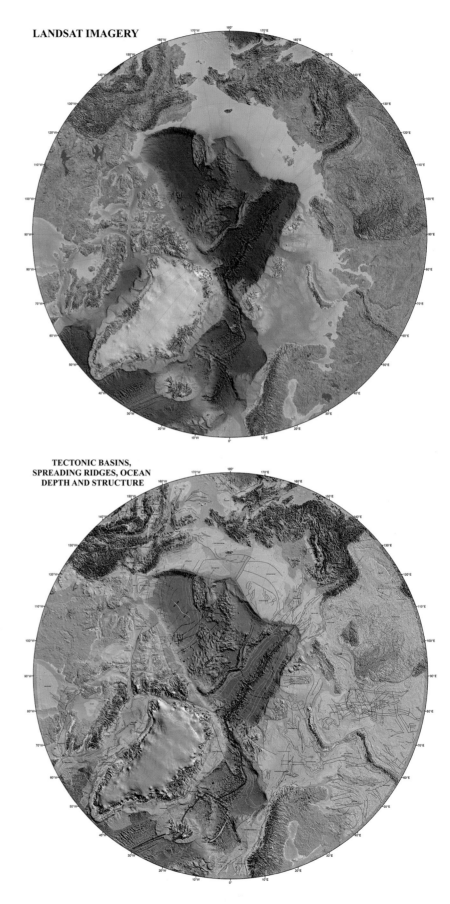

LANDSAT IMAGERY

TECTONIC BASINS, SPREADING RIDGES, OCEAN DEPTH AND STRUCTURE

ELEVATION

GRAVITY

**TECTONIC GEOLOGY,
SEAFLOOR AGE,
AND STRUCTURE**

**SEAFLOOR AGE,
LARGE IGNEOUS PROVINCES,
AND PASSIVE MARGINS**

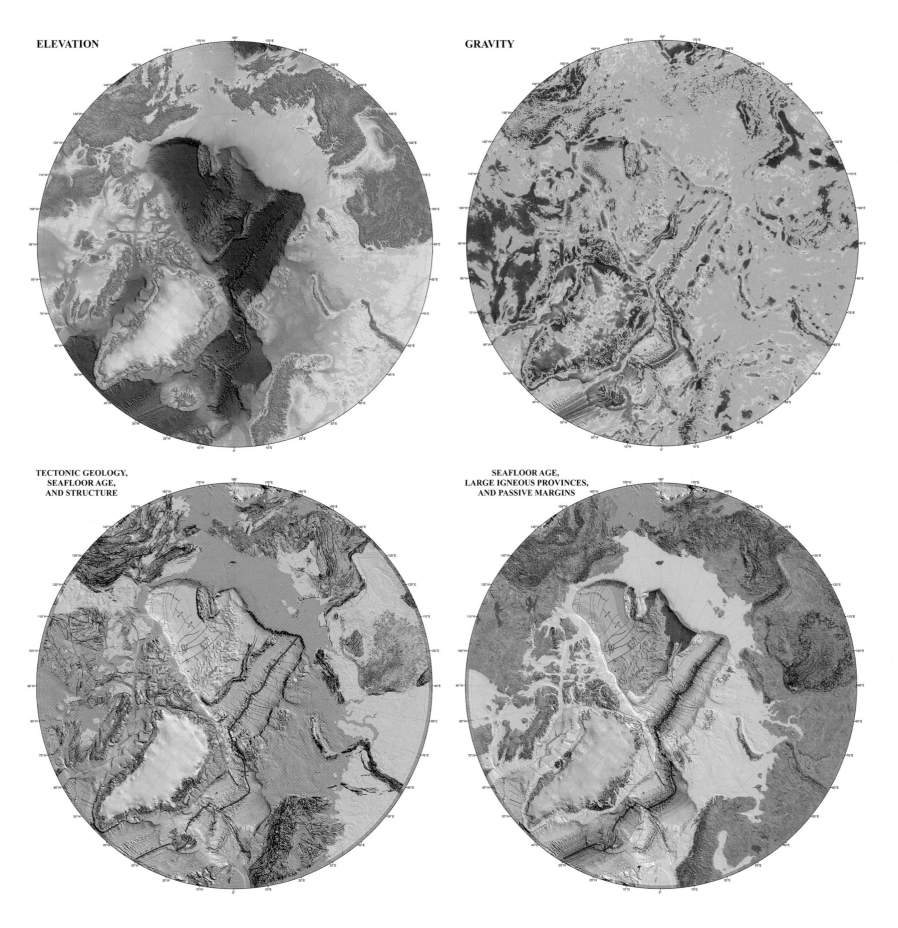

Atlanta Regional Map

GM Johnson & Associates Ltd.

Vancouver, British Columbia, Canada
By Guy Johnson

Contact

Guy Johnson
gjohnson@gmjohnsonmaps.com

Software

ArcGIS 10.1 for Desktop

Data Sources

Atlanta Regional Commission, Athens-Clarke County, Bartow County, Cobb County, Dekalb County, Floyd County, Fulton County, Georgia GIS Clearinghouse, Gordon County, Oconee County, US Census Bureau, US Geological Survey

This is an example of a folded local map for consumers by GM Johnson and Associates, Ltd., a Canadian company that specializes in street maps and atlas publishing for the North American retail market. The map of Atlanta, Georgia, and surrounding counties is a part of a series of over 450 published maps and atlases sold through retail channels. The *Atlanta Regional Map* took eight weeks to complete and the sister street map took six weeks to complete. The regional map is a generalization of the larger scale street map.

Courtesy of GM Johnson & Associates Ltd.

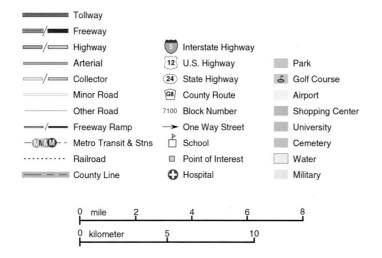

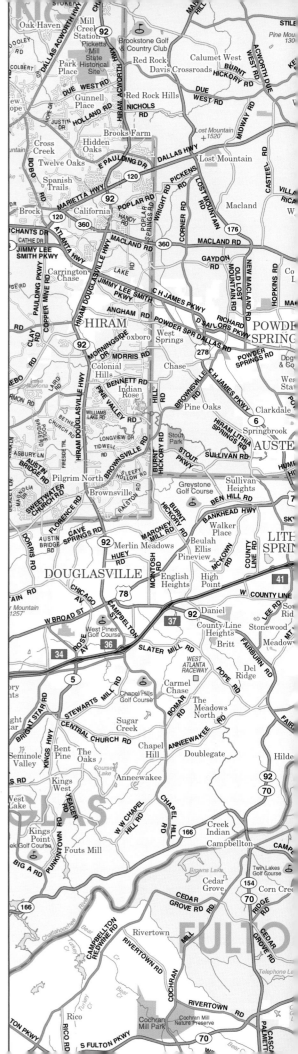

The Barents Map

Lantmäteriet, The Swedish Mapping, Cadastral and Land Registration Authority

Gävle, Sweden
By Parties of the Barents Geographic Data Base

Contact
Anders Sandin
anders.sandin@lm.se

Software
ArcGIS for Desktop

Data Source
National Geographic data

The Barents Region covers a vast portion of northern Europe—equal to the combined area of France, Spain, Germany, Italy, and the Netherlands— that extends over parts of Norway, Sweden, Finland, and Russia. National cadastre and mapping organizations and the regional cadastral offices of northwestern Russia have cooperated to produce a homogenous geographic database at the scale of 1:1 million covering the whole region. From the database, a number of derived products have been created, such as this map, generalized from the 1:1 million database. Users worldwide can assess the geographic and thematic information via the Internet.

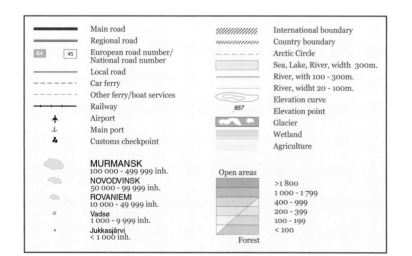

Main road		International boundary	
Regional road		Country boundary	
European road number/ National road number		Arctic Circle	
Local road		Sea, Lake, River, width 300m.	
Car ferry		River, with 100 - 300m.	
Other ferry/boat services		River, widht 20 - 100m.	
Railway		Elevation curve	
Airport		Elevation point	
Main port		Glacier	
Customs checkpoint		Wetland	
		Agriculture	

MURMANSK
100 000 - 499 999 inh.
NOVODVINSK
50 000 - 99 999 inh.
ROVANIEMI
10 000 - 49 999 inh.
Vadsø
1 000 - 9 999 inh.
Jukkasjärvi
< 1 000 inh.

Open areas
>1 800
1 000 - 1 799
400 - 999
200 - 399
100 - 199
< 100
Forest

Climatic Regions of the Czech Republic

Palacký University, Olomouc

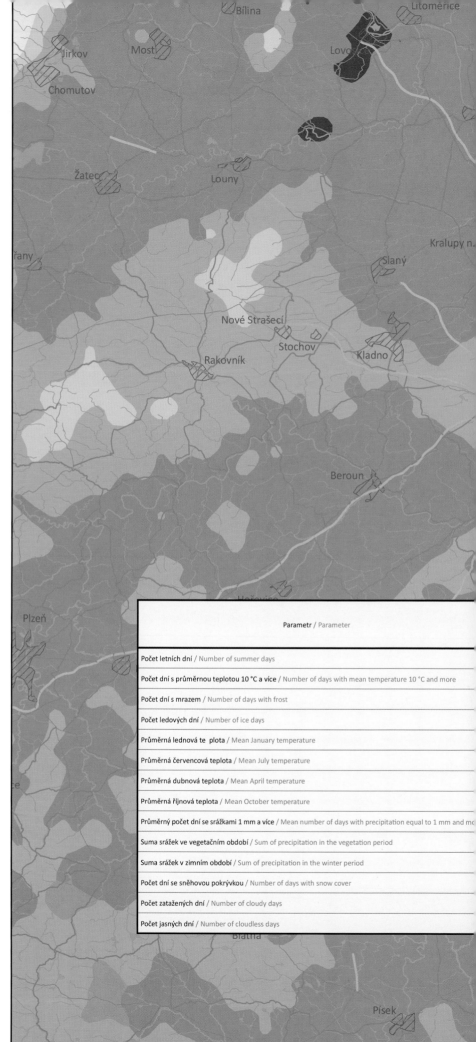

Olomouc, Olomouc Region, Czech Republic
By Alena Vondrakova, Ales Vavra, and Vit Vozenilek

Contact
Alena Vondrakova
alena.vondrakova@upol.cz

Software
ArcGIS 10.2 for Desktop, Adobe InDesign CS6

Data Source
Czech Hydrometeorological Institute

This map is based on the visualization of the Quitt climatology classification which divides a territory into climate regions according to complex characteristics in climatology. Some of the size classes are distinguished by a combination of selected meteorological elements. The study presented on this map is based on the approach used in the *Climate Atlas of Czechia* by Radim Tolasz (Prague: Czech Hydrometeorological Institute, 2007), which assesses the degree to which maps comply with the actual values of each meteorological characteristic. The study focused on defining climate regions based on Quitt classification and assessed the extent of uncertainty in the classification of locations and climatic areas of the classification scheme's original units.

Parametr / Parameter
Počet letních dní / Number of summer days
Počet dní s průměrnou teplotou 10 °C a více / Number of days with mean temperature 10 °C and more
Počet dní s mrazem / Number of days with frost
Počet ledových dní / Number of ice days
Průměrná lednová te plota / Mean January temperature
Průměrná červencová teplota / Mean July temperature
Průměrná dubnová teplota / Mean April temperature
Průměrná říjnová teplota / Mean October temperature
Průměrný počet dní se srážkami 1 mm a více / Mean number of days with precipitation equal to 1 mm and mo
Suma srážek ve vegetačním období / Sum of precipitation in the vegetation period
Suma srážek v zimním období / Sum of precipitation in the winter period
Počet dní se sněhovou pokrývkou / Number of days with snow cover
Počet zatažených dní / Number of cloudy days
Počet jasných dní / Number of cloudless days

	Klimatické charakteristiky chladných oblastí / Climate characteristics of cold regions						Klimatické charakteristiky mírně teplých oblastí / Climate characteristics of moderately warm regions											Klimatické charakteristiky teplých oblastí / Climate characteristics of warm regions				
	C2	C3	C4	C5	C6	C7	MW1	MW2	MW3	MW4	MW5	MW6	MW7	MW8	MW9	MW10	MW11	W1	W2	W3	W4	W5
	0–10	0–20	0–20	10–30	10–30	10–30	20–30	20–30	20–30	20–30	30–40	30–40	30–40	40–50	40–50	40–50	40–50	50–60	50–60	60–70	60–70	60–70
	0–80	80–120	80–120	100–120	120–140	120–140	120–140	140–160	120–140	140–160	140–160	140–160	140–160	140–160	140–160	140–160	140–160	160–170	160–170	170–180	170–180	>180
	160–180	160–180	160–180	140–160	140–160	140–160	160–180	110–130	130–160	110–130	130–140	140–160	110–130	130–140	110–130	110–130	110–130	120–130	100–110	110–120	100–110	90–100
	60–70	60–70	60–70	60–70	60–70	50–60	40–50	40–50	40–50	40–50	40–50	40–50	40–50	40–50	30–40	30–40	30–40	30–40	30–40	30–40	30–40	<30
	$-7--8$	$-7--8$	$-7--6$	$-5--6$	$-4--5$	$-3--4$	$-5--6$	$-3--4$	$-3--4$	$-2--3$	$-4--5$	$-5--6$	$-2--3$	$-4--5$	$-3--4$	$-2--3$	$-2--3$	$-3--5$	$-2--3$	$-3--4$	$-2--3$	$-1--2$
	10–12	12–14	12–14	14–15	14–15	15–16	15–16	16–17	16–17	16–17	16–17	16–17	16–17	17–18	17–18	17–18	17–18	17–19	18–19	19–20	19–20	19–20
	0–2	0–2	2–4	2–4	2–4	4–6	5–6	6–7	6–7	6–7	6–7	6–7	6–7	7–8	6–7	7–8	7–8	7–8	8–9	8–10	9–10	9–10
	2–4	2–4	4–5	5–6	5–6	6–7	6–7	6–7	6–7	6–7	6–7	6–7	7–8	7–8	7–8	7–8	7–8	7–9	7–9	8–9	9–10	9–10
	140–160	120–140	120–140	120–140	140–160	120–130	120–130	120–130	110–120	110–120	100–120	100–120	100–120	100–120	100–120	100–120	90–100	90–100	90–100	90–100	80–90	80–90
	700–900	600–700	600–700	500–600	600–700	500–600	500–600	450–500	350–450	350–450	350–450	450–500	400–450	400–450	400–450	400–450	350–400	350–400	350–400	350–400	300–350	300–350
	500–600	400–500	400–500	350–400	400–500	350–400	300–350	250–300	250–300	250–300	250–300	250–300	250–300	250–300	250–300	200–250	200–250	200–300	200–300	200–300	200–300	200–300
	160–200	140–160	140–160	120–140	120–140	100–120	100–120	80–100	60–100	60–80	60–100	80–100	60–80	60–80	60–80	50–60	50–60	50–80	40–50	50–60	40–50	<40
	130–150	140–150	130–150	140–150	140–150	150–160	120–150	150–160	120–150	150–160	120–150	120–150	120–150	120–150	120–150	120–150	120–150	120–140	120–140	110–120	110–120	<110
	30–40	30–40	30–40	30–40	40–50	40–50	40–50	40–50	40–50	40–50	50–60	40–50	40–50	40–50	40–50	40–50	40–50	40–50	40–50	50–60	50–60	50–60
	C2	C3	C4	C5	C6	C7	MW1	MW2	MW3	MW4	MW5	MW6	MW7	MW8	MW9	MW10	MW11	W1	W2	W3	W4	W5

Klimatické charakteristiky chladných oblastí
Climate characteristics of cold regions

Klimatické charakteristiky mírně teplých oblastí
Climate characteristics of moderately warm regions

Klimatické charakteristiky teplých oblastí
Climate characteristics of warm regions

Cascadia: A Great Green Land

Cascadia Institute

Eugene, Oregon, USA

By David McCloskey (Cascadia Institute) and Neil Allen (cartographer, Benchmark Maps)

Contact
David McCloskey
cascadia.institute@gmail.com

Software
ArcGIS for Desktop, Adobe Illustrator, Adobe Photoshop

Data Sources
SRTM30_Plus Global Topography, US Geological Survey Gap Analysis Program May 2011. US National Land Cover, British Columbia Biogeoclimatic Subzone/Variant Map, Global Land Ice Measurements from Space

Cascadia, a bioregion encompassing watersheds of the North Pacific Slope, is named for the cascades pouring down its slopes. Cascadia curves from the Pacific Ocean to the Rocky Mountains and Continental Divide. A bioregion is defined by its distinctive character and context. Its dynamic life is woven on many levels, including seafloor, coastline, terrain, climate, icefields, rivers, and vegetation. Cascadia has the largest interior temperate maritime conifer forests in the world. Shown here is a seamless cross-border view of the vegetative land cover of British Columbia and the United States. Based on the new integrative forest framework in the legend, each type lists the major species of that assemblage, both conifers and broadleafs, in order of prevalence.

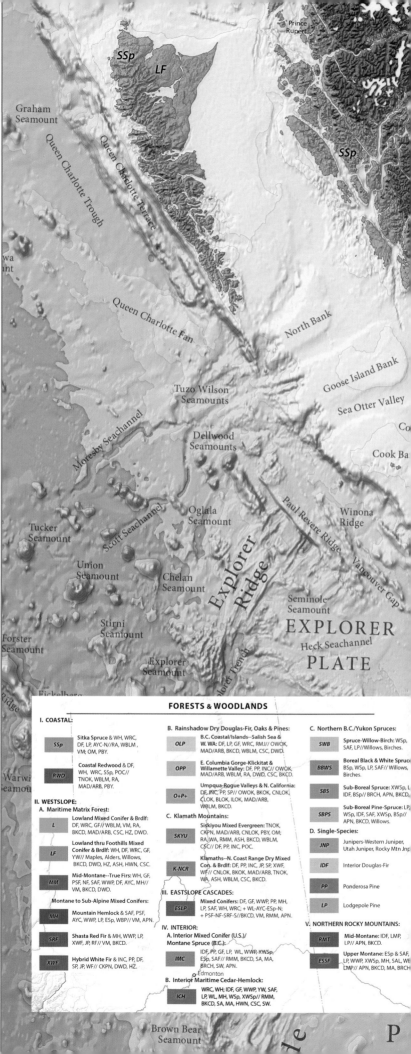

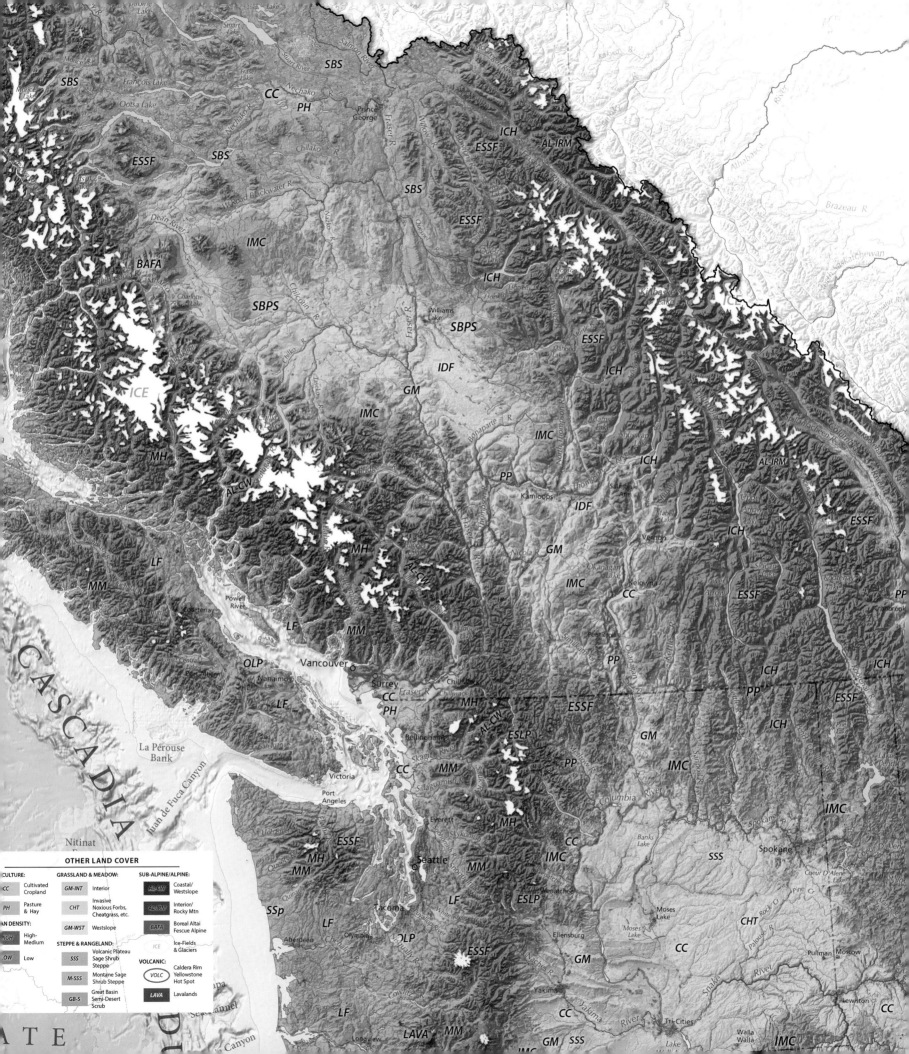

SBS

SBS

CC
PH

Prince George

SBS

ICH
AL-IRM

ESSF

Fraser R

ESSF

ESSF

SBS

IMC

BAFA

ICH

Williams Lake

SBPS

ICH

ESSF

SBPS

ICE

IDF

GM

ICH

IMC

IMC

PP

MH

ICH

AL-CW

Kamloops

IDF

ICH

GM

Vernon

AL-CW

MH

LF

Powell River

MM

CC

ESSF

Courtenay

Powell River

MM

LF

IMC

Kelowna

AL-IRM
IMC

Penticton

MM

PP

ICH

PP

LF

Vancouver

OLP

Surrey
CC
PH
Fraser R

Nanaimo

Chilliwack

MH
AL-CW

ESLP

GM

ICH

PP

Bellingham

MM

PP

ICH

Victoria

CC

MM

IMC

GM

Port Angeles

MH

IMC

ESSF

Everett

MH

CC

SSS

Spokane

MM

Seattle

MM

IMC

LF

LF

MH

MM

Wenatchee

ESLP

CHT

MM

Ellensburg

CC

Olympia

OLP

Moses Lake

SSp

LAVA

GM

SSS

Pullman Moscow

LF

CC

MM

GM
SSS

IMC

CASCADIA

La Pérouse Bank

Nitinat

Lewiston

Mallard Migration Widget: Tracking Fall Migration

Missouri Department of Conservation

Columbia, Missouri, USA

By Philip H. Marley, Tim Bixler, and Andy Raedeke

Contact

Philip H. Marley
philip.marley@mdc.mo.gov

Software

ArcGIS 10.0 for Desktop

Data Source

Missouri Mallard Observation Network Database

The Mallard Migration Observation Network (MMON) was established as part of a broader project to use GPS satellite telemetry to better understand mallard movements, distribution, and habitat use. The MMON participants, which include 150 wetland managers located throughout the Central and Mississippi Flyways, provide weekly fall migration reports of the mallard densities on a scale of 0–10 (10 = peak numbers) for the areas they manage. This subjective evaluation provides managers an opportunity to place the number of mallards they observe in context and serves as a measure of migration due to climatic conditions. The weekly migration maps are made available to the public through a web widget that is updated weekly with new migration maps during the fall migration season. Additionally, the data collected from the MMON was compared with the movements of mallards that are carrying the additional weight of a tracking transmitter to determine if there is any abnormal migration behavior.

Courtesy of Missouri Department of Conservation.

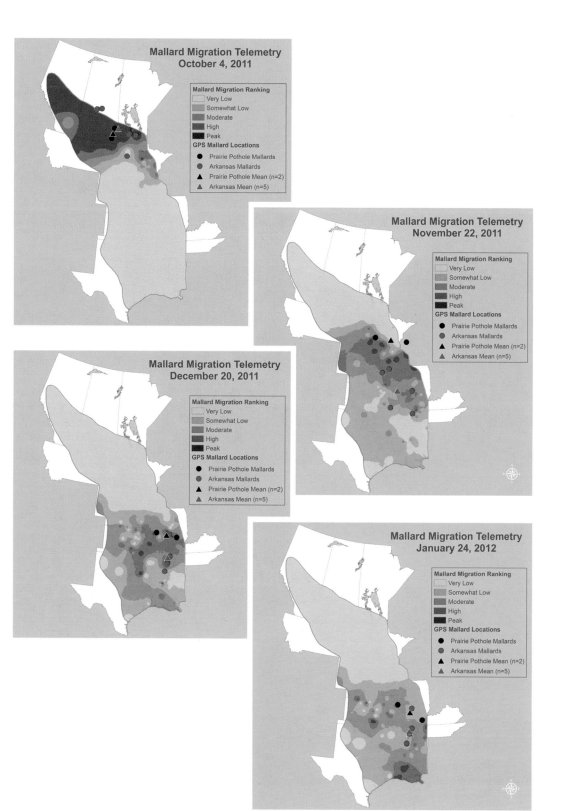

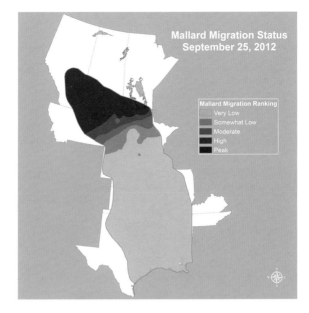

Mallard Migration Status
September 25, 2012

Mallard Migration Ranking
Very Low
Somewhat Low
Moderate
High
Peak

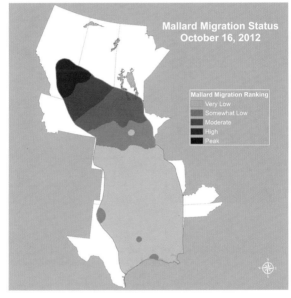

Mallard Migration Status
October 16, 2012

Mallard Migration Ranking
Very Low
Somewhat Low
Moderate
High
Peak

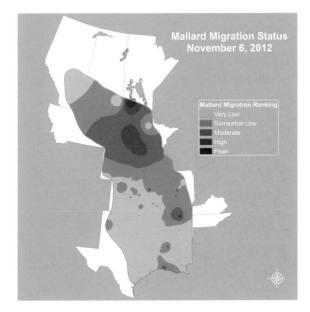

Mallard Migration Status
November 6, 2012

Mallard Migration Ranking
Very Low
Somewhat Low
Moderate
High
Peak

Mallard Migration Status
November 27, 2012

Mallard Migration Ranking
Very Low
Somewhat Low
Moderate
High
Peak

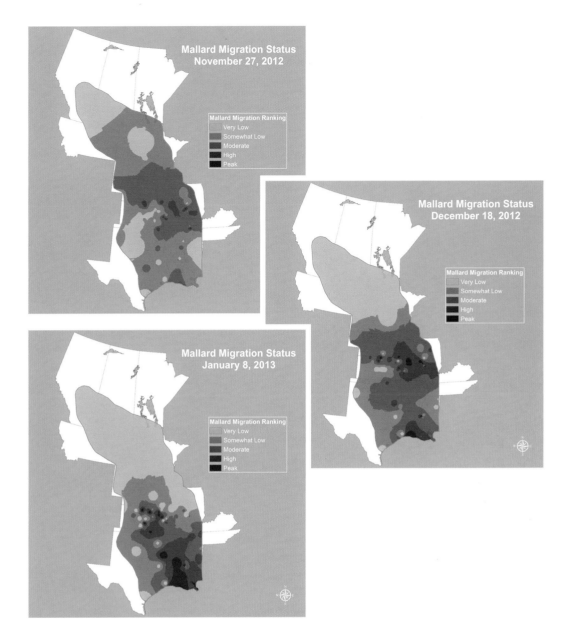

Mallard Migration Status
December 18, 2012

Mallard Migration Ranking
Very Low
Somewhat Low
Moderate
High
Peak

Mallard Migration Status
January 8, 2013

Mallard Migration Ranking
Very Low
Somewhat Low
Moderate
High
Peak

Widget Views
By Country

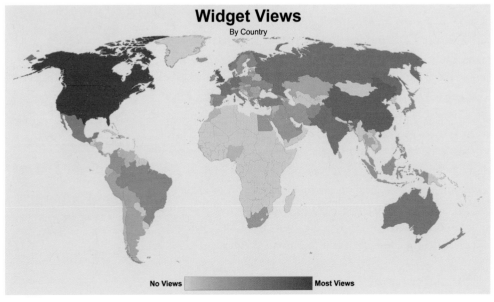

No Views   Most Views

25

Using the Condition of Arizona's Landscapes to Inform Conservation and Mitigation Priorities

The Nature Conservancy (TNC)

Tucson, Arizona, USA
By Marisa Guarinello, Marcos Robles, Dale Turner,
and Rob Marshall

Contact
Maria Guarinello
mguarinello@gmail.com

Software
ArcGIS 10.1 for Desktop

Data Sources
The Nature Conservancy, US Geological Survey, Arizona
Game and Fish Department, US Environmental Protection
Agency, Protected Areas Database of the United States,
Arizona State Land Department

The Nature Conservancy (TNC) developed a dataset that provides a powerful and easy way to understand the intensity and spatial distribution of human modifications to lands across Arizona. TNC used aerial imagery to classify more than 84,000 square-mile hexagons using two categorical values, one for the percent of land visibly used by humans and another identifying the dominant land-use type. Results show that over 50 percent of the state can be considered relatively intact and 17.5 percent of the state has been heavily used by humans. Knowing the location of large intact patches throughout the state informs TNC about areas that may be important for wildlife. This data helps TNC respond to development proposals by advocating that new development should avoid the largest of these intact lands. Pairing these patches with additional information helps TNC refine conservation priorities and recommend candidate locations for off-site mitigation actions. Further, combining data on human use with commonly used ecological system data provides an estimate of recent changes to each system due to human influences. Examining how intact and protected these systems are helps TNC identifies potential priorities for conservation, restoration, and mitigation.

Courtesy of The Nature Conservancy.

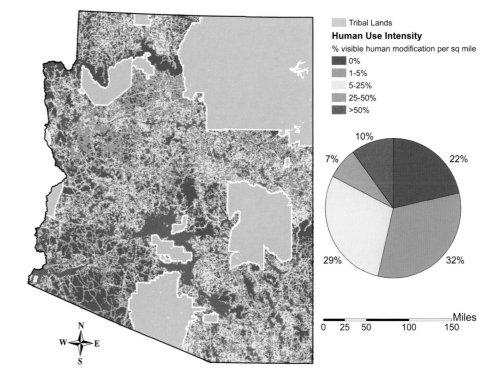

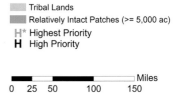

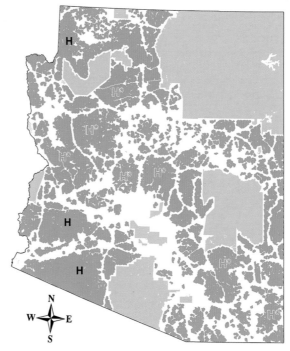

Evaluating the Value of Habitat Restoration Projects

Port of Seattle

Seattle, Washington, USA
By Randy Edwards, Tess Brandon, and Devlin Donnelly

Contact
Randy Edwards
Edwards.R@portseattle.org

Software
ArcGIS 10.2.2 for Desktop, ArcGIS Spatial Analyst and 3D Analyst, Adobe Creative Suite CS5, Microsoft Excel 2010, Microsoft Visio 2010

Data Source
Port of Seattle

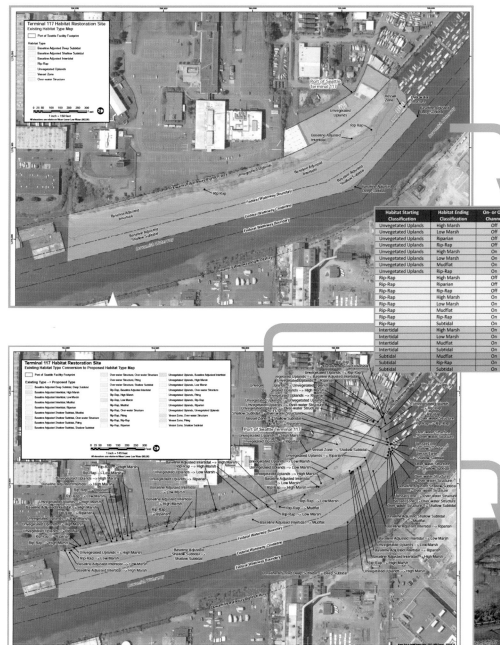

Habitat Starting Classification	Habitat Ending Classification	On- or Off-Channel	Area (sq-ft)	Area (acres)
Unvegetated Uplands	High Marsh	Off	65,614.06	1.50
Unvegetated Uplands	Low Marsh	Off	19,737.72	0.45
Unvegetated Uplands	Riparian	Off	84,885.35	1.94
Unvegetated Uplands	Rip-Rap	Off	760.08	0.01
Unvegetated Uplands	High Marsh	On	8,832.37	0.20
Unvegetated Uplands	Low Marsh	On	1,347.96	0.03
Unvegetated Uplands	Mudflat	On	1,304.21	0.03
Unvegetated Uplands	Rip-Rap	On	1,126.60	0.02
Rip-Rap	High Marsh	Off	131.80	0.00
Rip-Rap	Riparian	Off	4,530.62	0.10
Rip-Rap	Rip-Rap	Off	56.75	0.00
Rip-Rap	High Marsh	On	13,512.66	0.31
Rip-Rap	Low Marsh	On	22,687.24	0.52
Rip-Rap	Mudflat	On	21,233.33	0.48
Rip-Rap	Rip-Rap	On	4,657.33	0.10
Rip-Rap	Subtidal	On	1,498.13	0.03
Intertidal	High Marsh	On	709.77	0.01
Intertidal	Low Marsh	On	8,214.96	0.18
Intertidal	Mudflat	On	99,038.81	2.27
Intertidal	Subtidal	On	615.45	0.01
Subtidal	Mudflat	On	4,512.15	0.10
Subtidal	Rip-Rap	On	591.76	0.01
Subtidal	Subtidal	On	99,098.28	2.27

The US Environmental Protection Agency (EPA) identified the Terminal 117 site, located on the west bank of the Lower Duwamish Waterway in Seattle, for cleanup. The upland property, river bank and sediments, and parts of the adjacent yards and streets have high concentrations of PCBs (polychlorinated biphenyls), dioxin/furans, and other contaminants resulting from decades of industrial use. The EPA issued a plan to remove upland soil and river sediment to reduce PCBs and other contaminants to levels that will protect the river environment and reduce health risks to people. In addition to cleanup, trustee agencies involved in the process require that responsible parties provide compensation—typically fish and wildlife habitat restoration—in addition to cleanup to account for natural resource damages that have accrued over time. In the Pacific Northwest, the Habitat Equivalency Analysis (HEA) model is commonly used for resolving damage claims as well for determining compensatory mitigation required by the agencies overseeing the cleanup projects.

Courtesy of Port of Seattle, Seaport Environmental and Planning Program.

CONSERVATION AND SUSTAINABLE DEVELOPMENT

Mapping Urbanization in Dubai

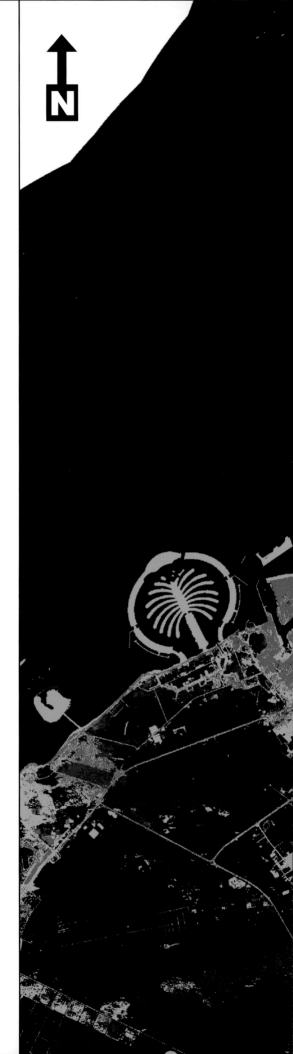

School of Design, University of Pennsylvania

Philadelphia, Pennsylvania, USA
By Shrobona Karkun

Contact
Shrobona Karkun
shrobona@design.upenn.edu

Software
ArcGIS 10.1 for Desktop, Adobe Illustrator CS5.5

Data Sources
Landsat 7 (2004) and Landsat 8 (2013), US Geological Survey Global Visualization Viewer

Dubai is the most populous city and emirate in the United Arab Emirates. In this map, satellite images and raster analysis show how urban development changed in Dubai between 2004 and 2013. GIS tools helped differentiate the natural environment (such as water and desert) from the built environment. Spatial analysis tools were used to determine how much urbanization took place, especially along the Dubai marina. This type of analysis helped gauge the extent of city expansion and sprawl that took place over nine years.

Courtesy of Penn Institute of Urban Research, Penn Design, University of Pennsylvania.

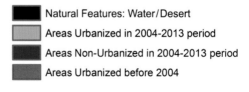

Natural Features: Water/Desert
Areas Urbanized in 2004-2013 period
Areas Non-Urbanized in 2004-2013 period
Areas Urbanized before 2004

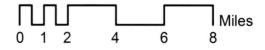

Miles
0 1 2   4   6   8

Toward a Sustainable Region—2031

The Regional Municipality of York

Newmarket, Ontario, Canada
By Geographic Information Services Branch,
commissioned by the former Long Range Strategic Planning Branch

Contact
John Houweling
John.Houweling@york.ca

Software
ArcGIS 10.0 for Desktop

Data Sources
The Regional Municipality of York, Queen's Printer for Ontario,
Ontario Ministry of Municipal Affairs and Housing

This map illustrates strategic policies adopted by the Regional Municipality of York's Council to help the region achieve its vision of sustainability by the year 2031. The map provides a visual accompaniment to the Regional Official Plan, which emphasizes strategic documents that will guide economic, environmental, and community-building decisions for managing growth. Located in the Greater Toronto and Hamilton Area, the region has a population of just over 1.1 million and is anticipated to reach 1.5 million residents by 2031. Woodlands cover more than 23 percent of the region and an extensive network of trails provides quality outdoor recreation for walking and cycling. York Region's natural beauty is complemented by a rich cultural heritage, including First Nations and Métis (mixed European and Native American ancestry) heritage sites. The region's agricultural industry produces a wide variety of locally grown fruit, vegetables, livestock, and dairy products. Sustainability is the lens through which York Region formulates, enhances, and implements policy. The vision for York Region in 2031 is to accommodate anticipated growth in a manner that ensures a sustainable natural environment, maintains a thriving economy, and creates healthy communities.

Courtesy of the Regional Municipality of York.

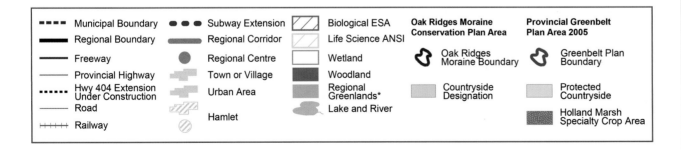

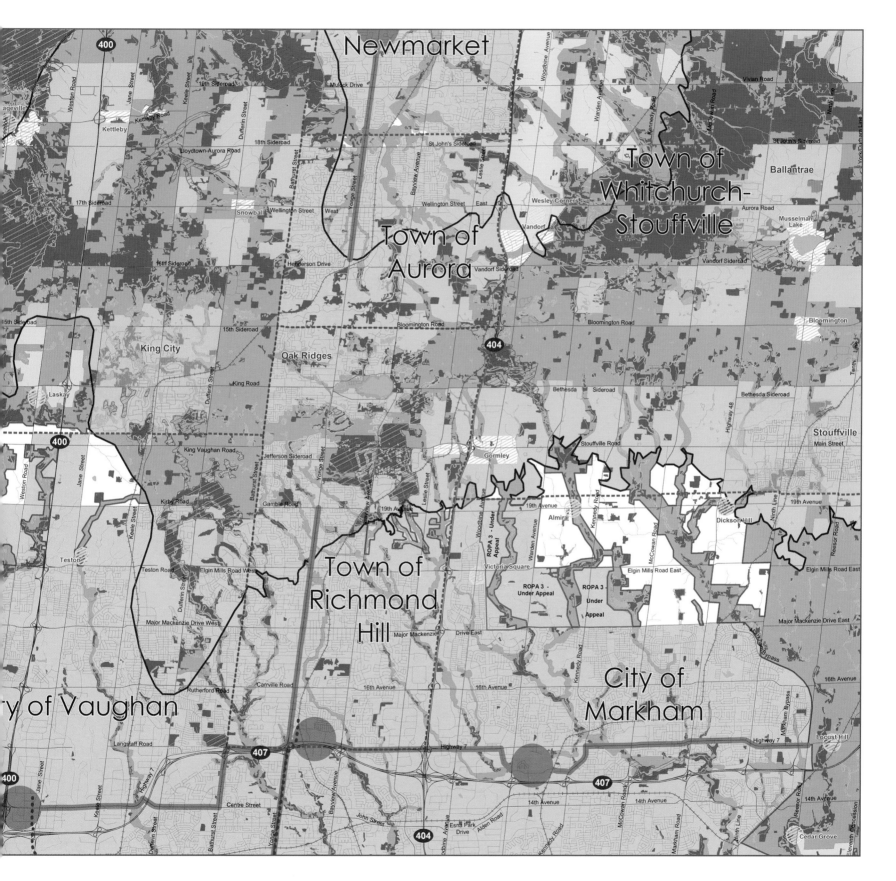

Newmarket

Town of Aurora

Town of Whitchurch-Stouffville

Ballantrae

Musselman Lake

King City

Oak Ridges

Bloomington

Stouffville

Laskay

Town of Richmond Hill

Gormley

Almira

Dickson Hill

ROPA 3 - Under Appeal

ROPA 3 - Under Appeal

ROPA 3 - Under Appeal

Victoria Square

Teston

City of Vaughan

City of Markham

Locust Hill

Cedar Grove

31

Mapping Important Natural Capital in Cambodia

Conservation International

Arlington, Virginia, USA
By Matt Gibb, Kellee Koenig, and Rachel Neugarten

Contact
Rachel Neugarten
rneugarten@conservation.org

Software
ArcGIS 10.2.1 for Desktop

Data Sources
Japanese International Cooperation Agency, LandScan, Open Development Cambodia, Critical Ecosystem Partnership Fund (2011), BirdLife International (2004), Plantlife International, Biodiversity and Protected Area Management Project, Waterworld version 2 (2014), Mekong Rivers Commission,

Natural capital is the biodiversity and ecosystems that produce the flow of goods, services, and other benefits to human well-being. Conservation International maps natural capital as a way to guide conservation priorities and measure the impact and effectiveness of conservation programs. These maps highlight areas of natural capital that are critically important for biodiversity, food security, and adaptation to climate change in Cambodia. The maps were produced by bringing together global and national-level datasets, conducting GIS analysis, and modeling key ecosystem services including provision of nontimber forest products and fresh water quantity, quality, and flow regulation. In the process of developing these maps, Conservation International conducted workshops with policy makers and stakeholders to validate the results found through spatial analysis. Working with stakeholders allows Conservation International to empower societies to responsibly and sustainably care for nature, global biodiversity, and the well-being of humanity.

Courtesy of Conservation International.

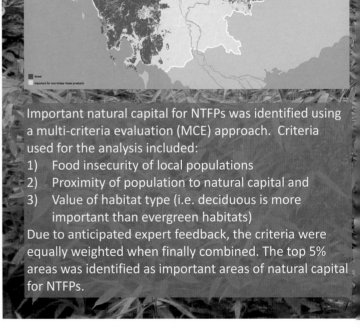

Natural Vegetation

Non-Timber Forest Products

Important natural capital for NTFPs was identified using a multi-criteria evaluation (MCE) approach. Criteria used for the analysis included:
1) Food insecurity of local populations
2) Proximity of population to natural capital and
3) Value of habitat type (i.e. deciduous is more important than evergreen habitats)
Due to anticipated expert feedback, the criteria were equally weighted when finally combined. The top 5% areas was identified as important areas of natural capital for NTFPs.

Natural capital is the stock of ecosystems and biodiversity that provide a flow of goods and services that support human well-being. Identifying important natural capital through GIS analysis and stakeholder engagement can ensure that the most important areas of natural capital are conserved and sustained. The ecosystem services identified as most important in Cambodia include:

- carbon storage
- non-timber forest products
- climate adaptation
- biodiversity
- fresh water flows
- fisheries
- flood regulation
- rice irrigation
- cultural/tourism services

Only three of these ecosystem services are highlighted here, but our analysis help us identify which natural ecosystems (left) are the most important natural capital (right), and if they are sustained through protection.

Important Natural Capital

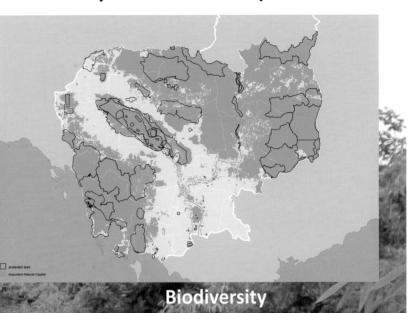

protected area
Important Natural Capital

Climate Adaptation

deep pool and 1km surrounding area
important for freshwater quantity and regulation
climate change with hydropower

Identifying natural capital for climate adaptation depends on results from analyses relating to freshwater and food security. This analysis combined four identifiers:
1) freshwater regulation
2) freshwater quantity
3) important deep pool fish sites and
4) areas facing greatest threats to hydropower and climate change

Biodiversity

important for biodiversity

Important natural capital for biodiversity was defined as habitats harboring threatened and protected species and ecosystems. The data used to identify these areas were:
1)   Key biodiversity areas (KBAs)
2)   Important birding areas (IBAs) and
3)   Presence of vulnerable, endangered, and critically endangered species
In order to ensure that the important areas were intact ecosystems, we used the most current forest cover data to extract any deforested land from these identifiers.

Site Suitability for Communication Towers

SRI International

Menlo Park, California, USA
By Reina Kahn

Contact
Reina Kahn
reina.kahn@sri.com

Software
ArcGIS 10.0 for Desktop

Data Sources
US Geological Survey, US Census Bureau
TIGER data

The eXportable Combat Training Capability (XCTC) program provides live tracking training for the US Army National Guard. To optimize communications, relay towers must be erected throughout the military installation. A suitable location requires high elevation relative to the rest of the terrain, flat land, and proximity to a drivable road. These requirements were weighted, based on recommendations of subject matter experts, to produce a suitability surface that ranked every area of the military installation. The process helped determine the location of communication tower placement and minimized field survey time.

Courtesy of SRI International.

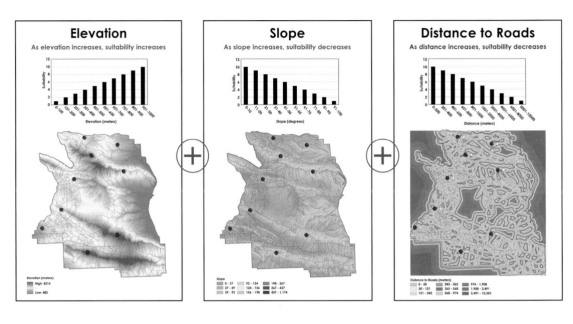

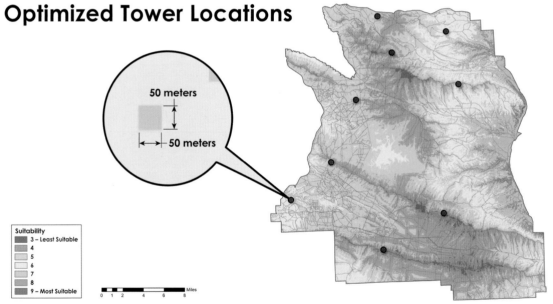

Optimized Tower Locations

Line-of-Sight Analysis for Range Development

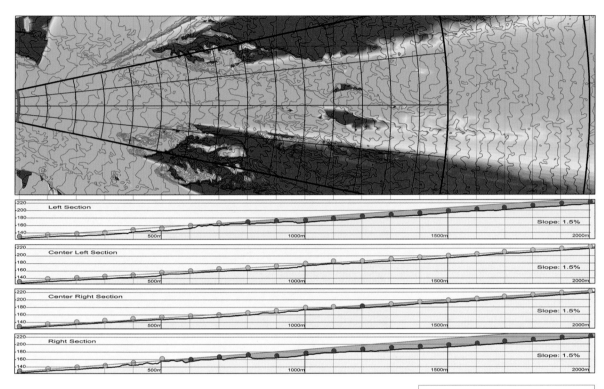

Tower Level Firing Position Viewshed Analysis

Ground Level Firing Position Viewshed Analysis

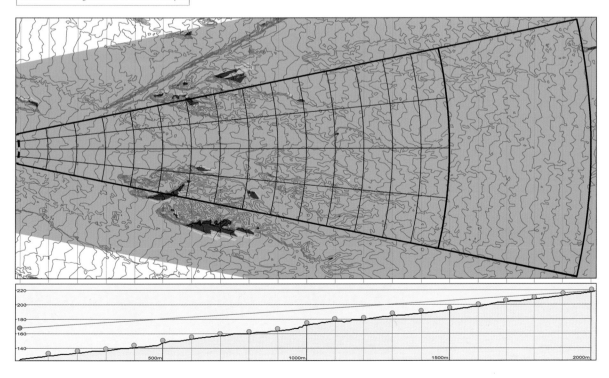

KTU+A

San Diego, California, USA
By Mike Limburg

Contact
Mike Limburg
mikel@ktua.com

Software
ArcGIS 10.1 for Desktop, ArcGIS 3D Analyst, AutoCAD 2014, Adobe Illustrator CC, Adobe Photoshop CC

Data Source
US Geological Survey National Elevation Dataset

These maps show a quick and easy method for presenting line of sight analysis to evaluate site options for a rifle range. The accompanying graphics show the viability of site options. The target visibility from the firing line is indicated with green and areas of visual obstruction are red. Target visibility is presented in both plan and section views. The increase in the visibility from the top image to the bottom image is the result of raising the firing position to a 40-foot tower. Areas in yellow are visible to some firing line positions but not others. The maps illustrate the extent to which topography affects target visibility. The sections that accompany each plan view diagram help illustrate specific target locations and the magnitude of the obstruction to help with decision making.

Courtesy of KTU+A.

Airfield Approach/Departure Surface Obstruction Study

US Army Corps of Engineers (USACE) and Fugro Consult GmbH

Munich, Bavaria, Germany
By Thomas Rodehaver

Contact
Thomas Rodehaver
thomas.r.rodehaver@usace.army.mil

Software
ArcGIS 10.2 for Desktop

Data Source
US Army Corps of Engineers

The US Army Corps of Engineers, Europe District, provides planning and design services for Department of Defense organizations within Europe and Africa in support of military forces. The USACE and its mapping partner, Furgo Consult GmbH, deliver site data in raster and vector formats through aerial imagery and ground survey. To create these maps, an Air Force Base in Europe and surrounding townships were flown over in thirty-two passes and covered with 2,572 images. The lidar was collected during the same pass coverage. Lidar data was used to build the 3D structures which included roof pitch/slopes and overhang.

Courtesy of US Army Corps of Engineers and Fugro Consult GmbH.

ArcMap: Overall Area Overview with new imagery and 3D Structures Overlayed on ArcGIS Online Imagery Basemap

ArcGlobe: 3D Perspective using ArcGlobe with Imagery and 3D Structures

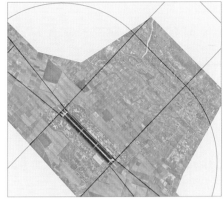

ArcScene: Overview with Flight Approach/ Departure Conical Surfaces

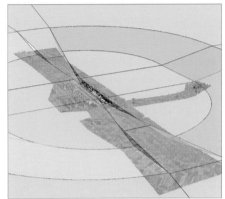

ArcScene: 3D View with Flight Approach/ Departure Conical Surfaces

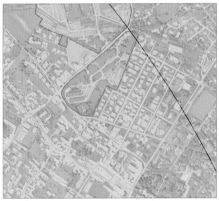

ArcScene: Area with Identified Conical Surface Arc

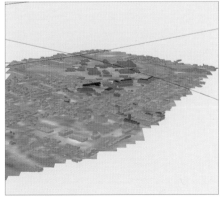

ArcScene: 3D View of Area with Flight Surface Obstruction

37

Identification of Potential Larval Habitat for *Culex* Mosquitoes

University of Illinois at Urbana-Champaign

Champaign, Illinois, USA
By Trisha Rentschler, Phong Le, and Surendra Karki

Contact
Trisha Rentschler
trentsch@illinois.edu

Software
ArcGIS 10.2.1 for Desktop

Data Sources
Centers for Disease Control and Prevention, National Resources Conservation Service, Illinois State Geological Survey; Machault, V., C. Vignolles, F. Borchi, P. Vounatsou, F. Pages, S. Briolant, J.P. Lacaux, and C. Rogier,, 2011. "The Use of Remotely Sensed Environmental Data in the Study of Malaria." *Geospatial Health* 5 (2):151–68.

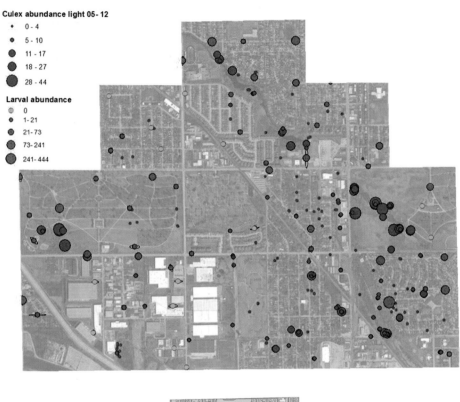

When weather conditions are ideal for mosquito reproduction, the surrounding environment has the potential for a decrease in healthy vegetation, leading residents to water their lawns. This additional water could potentially alter the mosquitoes breeding habits and habitat within the drainage basins and other water collection sites. Researchers at the University of Illinois Urbana-Champaign conducted a study to help them identify potential natural water larval habitat and lower-elevation areas that can persistently hold water. Using hydrological modeling, they identified and mapped drainage basins and water collection sites using the MATLAB computing environment and ArcGIS software. To further understand this complex ecological system, they collected mosquito larvae and used spatial modeling to determine how adding nonnatural water and rainfall to the surrounding environment affects the abundance of mosquito habitat and larvae.

Courtesy of University of Illinois at Urbana-Champaign.

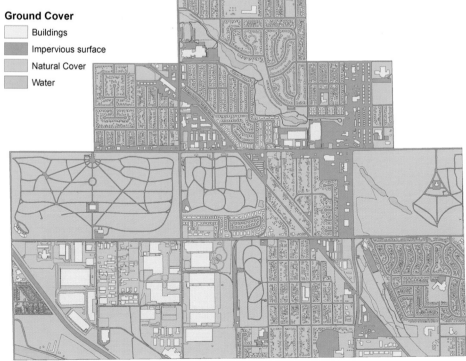

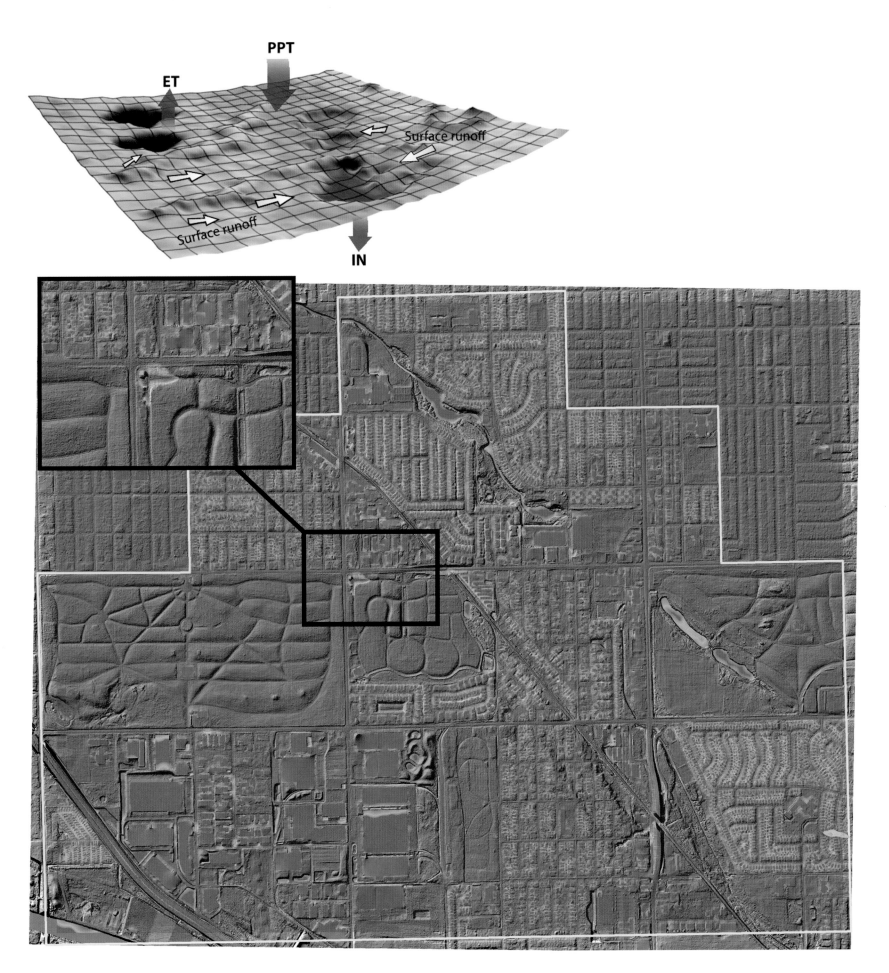

Using Lidar and Spectral Imagery as Land Management Tools

US Army Europe Sustainable Range Program, Parsons, IABG

Hohenfels, Bavaria, Germany
By Sylvia Guenther, Elke Kraetzschmar, and Steve Bowley

Contact
Sylvia Guenther
guenthers@iabg.de

Software
ArcGIS 10.1 for Desktop, Trimble eCognition 8.9

Data Sources
US Army Europe Sustainable Range Program lidar, aerial imagery, VHR WorldView-2

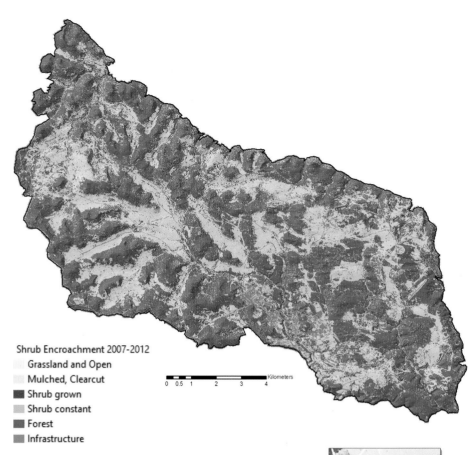

Shrub Encroachment 2007-2012
- Grassland and Open
- Mulched, Clearcut
- Shrub grown
- Shrub constant
- Forest
- Infrastructure

The Joint Multinational Readiness Center (JMRC) Hohenfels is a 161-square kilometer US Army training area located in Bavaria, Germany. The training area has been extensively used for maneuver exercises for over sixty years. Approximately 92 percent of the area has been designated European Union Natura 2000 Flora and Fauna Habitat (FFH) and bird habitat protection area. A dramatic expansion of populations of invasive species, such as the blackthorn shrub, has challenged land managers. The invasive species significantly reduce maneuver training capability and threaten the preservation of hundreds of protected species. Controlling the shrub encroachment is essential to protect Nature 2000 areas and to ensure optimal accessibility, availability, and capability of training land. IABG and Parsons performed a vegetation classification and change-detection analysis at JMRC Hohenfels between 2007 and 2012 to detect encroachment patterns. The approach used multitemporal lidar data and spectral imagery (aerial and satellite) combined with object-based image analysis. This proved to be a reliable, fast, accurate, and repeatable method of tracking shrub encroachment. The classification results enable range planners, trainers, and natural resource managers to identify, quantify, and prioritize areas for management activities.

Courtesy of Parsons and IABG on behalf of the US Army Europe.

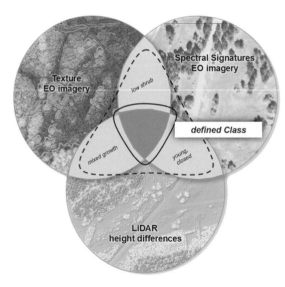

2007 \ 2012	Grassland and Open	Low Status	Medium and High Status	Forest
Grassland and Open	Grassland and Open	Shrub new	Shrub new	LiDARerror
Low Status	Mulched	Shrub constant	Shrub grown	LiDARerror
Medium and High Status	Mulched	Shrub constant	Shrub constant	Forest new
Forest	Clearcut	Clearcut	Forest constant	Forest constant

Change (%)	
no change	88.7
moderate change	3.8
significant change	4.2
infrastructure	3.3

Characterizing Polychlorinated Biphenyl (PCB) Sediment Contamination Using GIS

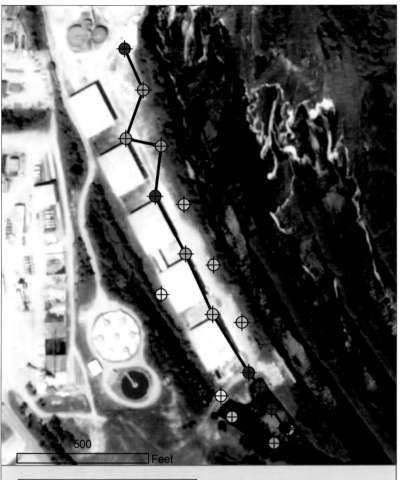

Area of Concern — PCB Contamination

Monitoring Well-PCBs (mg/kg)
0 - 1   1 - 10   10 - 50   50 - 500   500 - 1003

Transect (for cross section perspective)

N

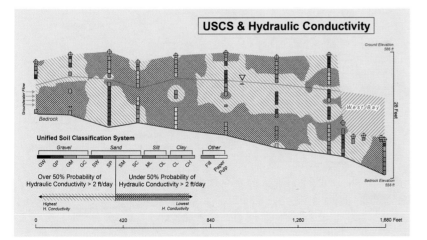

USCS & Hydraulic Conductivity

Groundwater Flow

Bedrock

West Bay

Ground Elevation 586 ft

28 Feet

Bedrock Elevation 558 ft

Unified Soil Classification System

Gravel	Sand	Silt	Clay	Other
GW GP GM GC	SW SP SM SC	ML CL	CL CH	Fill Paper Pulp

Over 50% Probability of Hydraulic Conductivity > 2 ft/day   Under 50% Probability of Hydraulic Conductivity > 2 ft/day

Highest H. Conductivity   Lowest H. Conductivity

0   420   840   1,260   1,680 Feet

CSS-Dynamac and US Environmental Protection Agency (EPA)

Cincinnati, Ohio, USA

By Alexander Hall, Marc Mills, Kyle Fetters, and Brian Crone

Contact
Alexander Hall
hall.alexander@epa.gov

Software
ArcGIS 10.0 for Desktop

Data Source
US Environmental Protection Agency field data

Polychlorinated biphenyl (PCB) sediment contamination poses great risks to human health and the environment. To assess exposure risk and identify plans for remediation, it is important to accurately characterize site conditions for PCB contamination. However, the way PCBs move to sediments through soils and groundwater, sometimes in sporadic routes, poses a significant challenge when evaluating migration using discrete soil sample data. Soil class, hydraulic conductivity, groundwater flow, presence of bedrock, soil carbon content, and lateral and horizontal pathway perspectives were all taken into account in this study for a more accurate representation of PCB sediment contamination. By finding significant relationships between contamination and contamination indicators, such as industrial waste and organic carbon-rich soils that PCBs readily bond to, researchers were able to more confidently define potentially contaminated areas outside of discrete soil sample locations.

Courtesy of CSS-Dynamac and US Environmental Protection Agency.

ENVIRONMENTAL MANAGEMENT
Tacoma Smelter Plume

Washington Department of Ecology

Lacey, Washington, USA
By Ian Mooser

Contact
Ian Mooser
imooser@yahoo.com

Software
ArcGIS 10.1 for Desktop, Adobe Illustrator CC

Data Sources
US Geological Survey, US Census Bureau, Washington Department
of Transportation

For almost 100 years, the Asarco Company operated a copper smelter
in the Tacoma area. Air pollution from the smelter settled on the surface
soil covering more than 1,000 square miles of the Puget Sound basin. The
Tacoma smelter specialized in processing ores with high arsenic levels, and
a large smokestack dispersed these as air emissions. Beginning in 1999, the
Washington Department of Ecology did a series of soil-sampling studies in
King, Pierce, Thurston, and Kitsap Counties, which helped to define the
nature and extent of the plume. Since then, the department has collected
many additional samples to create this analysis/map, which predicted arsenic
levels by census block group in 2014. This map shows that the two predominant factors regarding the pattern of the plume are distance and direction
from the former smelter. Contamination levels decrease with distance, and
they are higher along the dominant north-northeast and south-southwest
wind directions. The Department of Ecology uses this map for public and
local jurisdiction outreach as well as for planning purposes. This outreach is in
conjunction with ecology programs that offer soil sampling and soil cleanup
in the most contaminated areas of the plume.

Courtesy of Washington Department of Ecology.

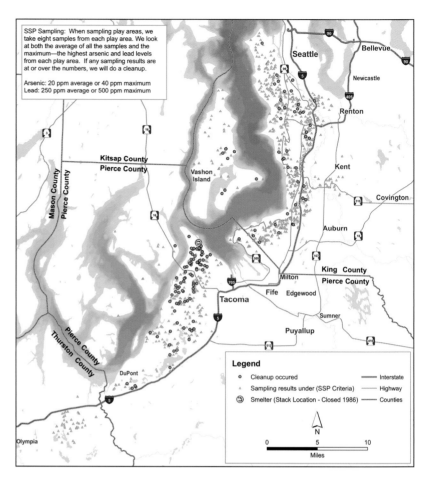

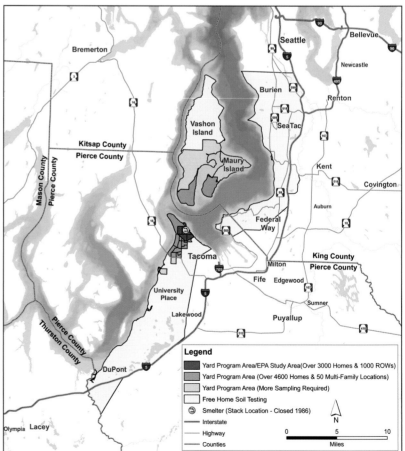

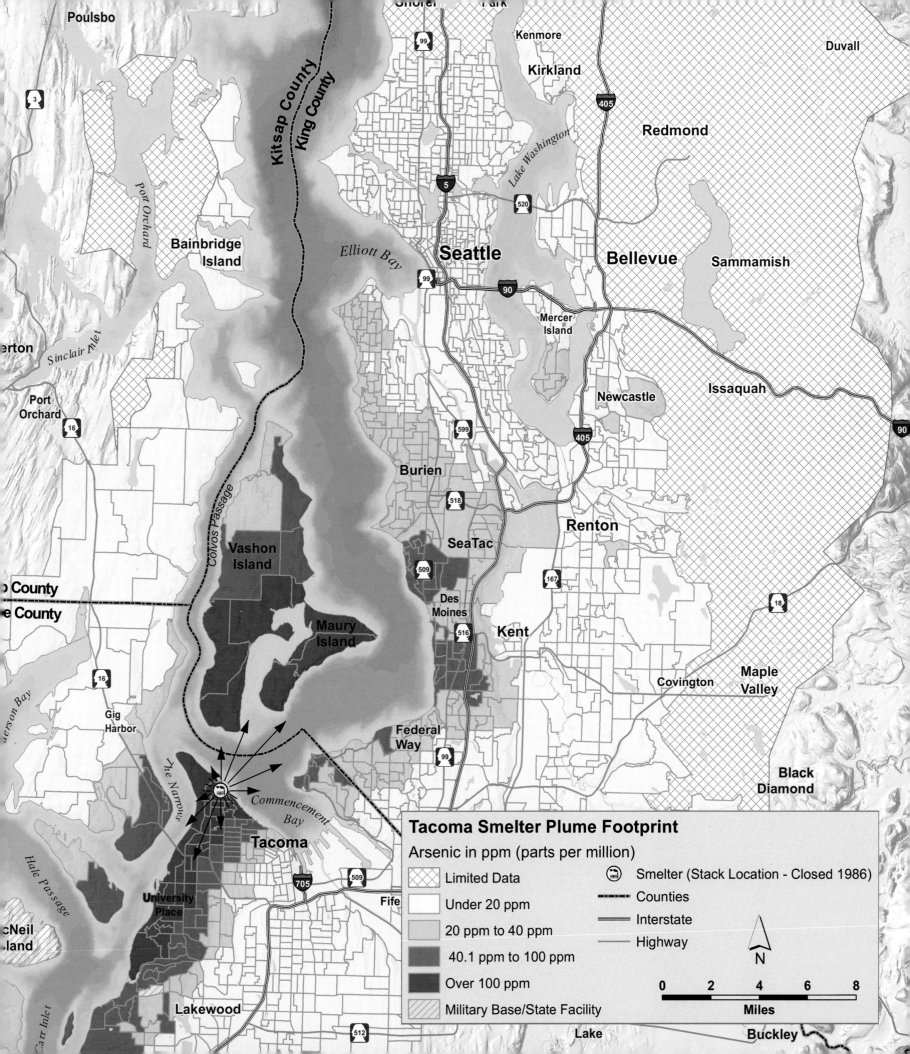

Tacoma Smelter Plume Footprint

Arsenic in ppm (parts per million)

Limited Data

Under 20 ppm

20 ppm to 40 ppm

40.1 ppm to 100 ppm

Over 100 ppm

Military Base/State Facility

Smelter (Stack Location - Closed 1986)

Counties

Interstate

Highway

N

0   2   4   6   8
Miles

Poulsbo

Kenmore

Kirkland

Duvall

Redmond

Bainbridge
Island

Kitsap County
King County

Elliott Bay

Seattle

Bellevue

Sammamish

Port Orchard

Mercer
Island

Sinclair Inlet

...erton

Port
Orchard

Newcastle

Issaquah

Colvos Passage

Burien

...o County
...e County

Vashon
Island

SeaTac

Renton

...on Bay

Maury
Island

Des
Moines

Kent

Maple
Valley

Covington

Gig
Harbor

The Narrows

Federal
Way

Black
Diamond

Commencement
Bay

Hale Passage

Tacoma

University
Place

Fife

...cNeil
...land

Carr Inlet

Lakewood

Lake

Buckley

America's Last Remaining Places of Natural Quiet

Stone Environmental and Noise Pollution Clearinghouse

Montpelier, Vermont, USA
By Katie Budreski (Stone Environmental) and Les Blomberg (Noise Pollution Clearinghouse)

Contact
Katie Budreski
katiebudreski@gmail.com

Software
ArcGIS 10.1 for Desktop

Data Sources
University of Nebraska and Stone Environmental, Inc., US Department of Transportation, US Census Bureau, Esri StreetMap 2007

Over the past 100 years, areas of natural quiet in the continental United States have shrunk from a predicted 75 percent to 2 percent. That 2 percent is in danger due to the rapid growth of motorized ground and air travel. In 1900, rail, steamboat, and roadway noise in populated areas influenced the soundscape. The first modern automobile was released by 1886, but the real birth of what we know today as modern transportation was in 1908 with the release of the Ford Model T.

Natural quiet areas in 1900 were estimated using historic railroads, navigable waterways, and historic population centers greater than 5,000 people per square mile. Over the past century, the automobile has made its way into nearly every household, and roads have been paved and widened to accommodate faster and louder travel. Air travel has also contributed to the modern soundscape. Current natural quiet areas were estimated using railroads, roads, and domestic air traffic for a single point in time. Although some solitude is possible in any of the dark green areas on the map, some spots are more likely to be free from road-, rail-, and air-related noise for extended periods. These areas mostly consist of protected federal lands and designated wilderness. The analysis accounted for some air traffic, but did not include international or military flight paths. The thirteen areas highlighted in purple are those patches without roads and rails that are farthest from all known flight paths.

Courtesy of Stone Environmental and Noise Pollution Clearinghouse.

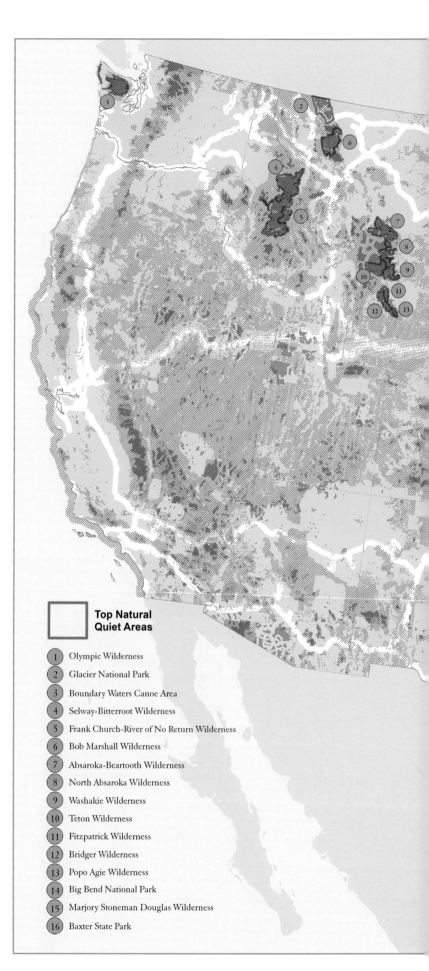

Top Natural Quiet Areas

1. Olympic Wilderness
2. Glacier National Park
3. Boundary Waters Canoe Area
4. Selway-Bitterroot Wilderness
5. Frank Church-River of No Return Wilderness
6. Bob Marshall Wilderness
7. Absaroka-Beartooth Wilderness
8. North Absaroka Wilderness
9. Washakie Wilderness
10. Teton Wilderness
11. Fitzpatrick Wilderness
12. Bridger Wilderness
13. Popo Agie Wilderness
14. Big Bend National Park
15. Marjory Stoneman Douglas Wilderness
16. Baxter State Park

Historic Areas of Natural Quiet circa 1900

Current Areas of Natural Quiet

Federally Protected Land

Designated Wilderness Areas

45

Modeling Large Stochastic Wildfires

US Department of Agriculture (USDA) Forest Service

Corvallis, Oregon, USA
By Cole Belongie and Ray Davis

Contact
Cole Belongie
cbelongie@fs.fed.us

Software
ArcGIS 10.1 for Desktop, Geospatial Modeling Environment 0.7.2.1

Data Sources
USDA Forest Service, Esri

Predicting the potential amount and location of critical habitat loss caused by wild-fire is difficult and can be highly debatable. This project details one method that was devised to predict the potential loss of habitat in the Northwest Forest Plan Area due to stochastic (random) large wildfires for the next five decades. This modeling focused on wildfires over 1,000 acres that routinely cause the most significant habitat loss. The Forest Service used the wildfire suitability model developed for the 15-year monitoring report for the Northwest Forest Plan. The Geospatial Modeling Environment software generated the one hundred random points. To match the amount and frequency of areas reburned, points were buffered by 5 kilometers (3.1 miles) for each decade which established an exclusion zone for the next decade's randomly generated points. After completing the analysis, both the modeled and observed wildfire severity classes and acres burned showed strong relationships over roughly the same time period.

Courtesy of USDA Forest Service.

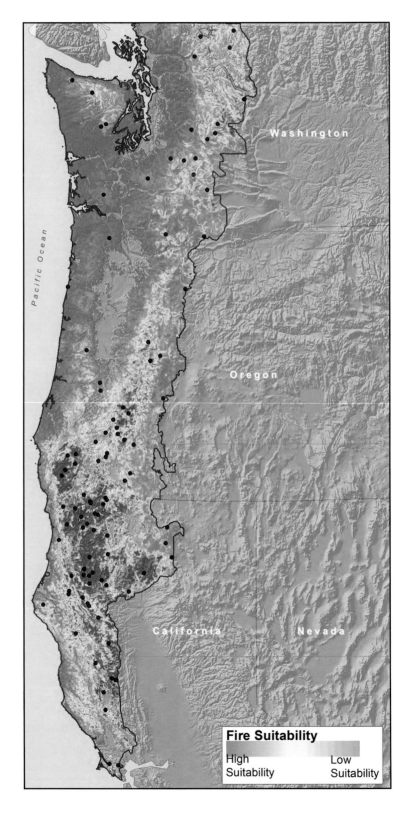

Example of wildfire probability surface and 1 decade of randomly generated points.

Fire Suitability

High Suitability — Low Suitability

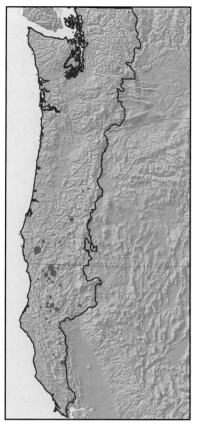

*Decade 1 Randomly Modeled
Wildfire Footprints*

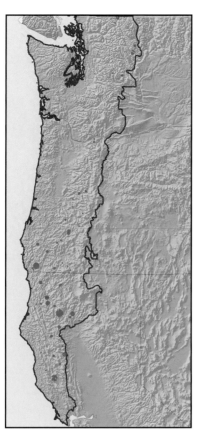

*Decade 2 Randomly Modeled
Wildfire Footprints*

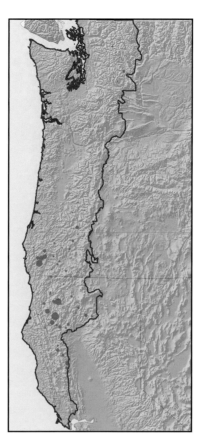

*Decade 3 Randomly Modeled
Wildfire Footprints*

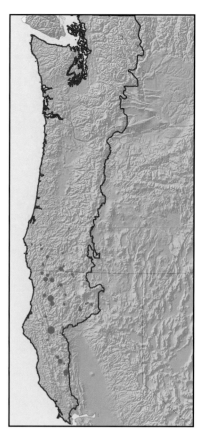

*Decade 4 Randomly Modeled
Wildfire Footprints*

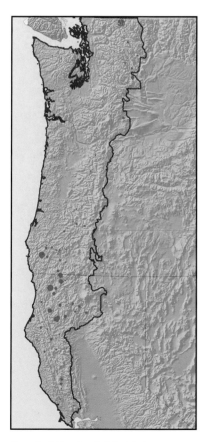

*Decade 5 Randomly Modeled
Wildfire Footprints*

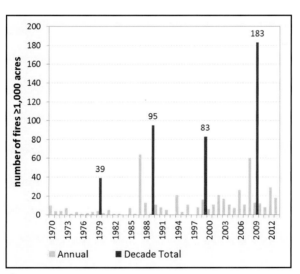

*Northwest Forest Plan monitoring
data for wildfires that have
exceeded 1,000 acres in size
(1970-2013). Labels above bars
represent decadal averages.*

Type of Burn	Real (acres)	Modeled (acres)
1 Burn (only burned in one decade)	3,400,600	3,302,200
2 Burns (burned in two separate decades)	909,200	959,300
3 Burns (burned in three separate decades)	68,900	99,000
Total area burned (includes additional acres for reburns)	4,378,700	4,360,500

Northwest Forest Plan Area summary of observed and modeled wildfire area burned.

Submersed Aquatic Vegetation Trends in Chesapeake Bay: 1984–2013

Virginia Institute of Marine Science (VIMS)

Gloucester Point, Virginia, USA

By Erica R. Smith, Anna K. Kenne, L. Nagey, Robert J. Orth, Jennifer R. Whiting, and David J. Wilcox

Contact
Erica R. Smith
ericasmi@vims.edu

Software
ArcGIS 10.1 for Desktop, Adobe Illustrator CS5

Data Source
Virginia Institute of Marine Science Submerged Aquatic Vegetation program annual monitoring data

The Virginia Institute of Marine Science (VIMS) conducts an annual aerial survey of submersed aquatic vegetation (SAV) in Chesapeake Bay and the adjacent Delmarva Coastal Bays. Seventeen species of SAV can be found throughout Chesapeake Bay. Their distribution is limited by salinity. Because of their sensitivity to water quality changes, SAV are used by resource managers as a sentinel group to monitor the condition of Chesapeake Bay annually. The data that VIMS provides is used by federal and state agencies to direct and evaluate restoration and management efforts in the Chesapeake Bay region. Four representative regions were selected within salinity zones defined for the bay area and featured on this map. Current SAV coverage for almost all zones in Chesapeake Bay remain below established restoration targets, indicating that SAV abundance and associated ecosystem services are currently limited by continued poor water quality. Unseasonably warm temperatures in 2006 and 2011 resulted in large die-back events in the higher salinity zones in Chesapeake Bay. In contrast, SAV has rebounded in the higher-salinity southern Delmarva Coastal Bays due to sustained restoration efforts assisted by better water quality conditions and lower water temperatures in these coastal bays.

Courtesy of Virginia Institute of Marine Science, College of William & Mary.

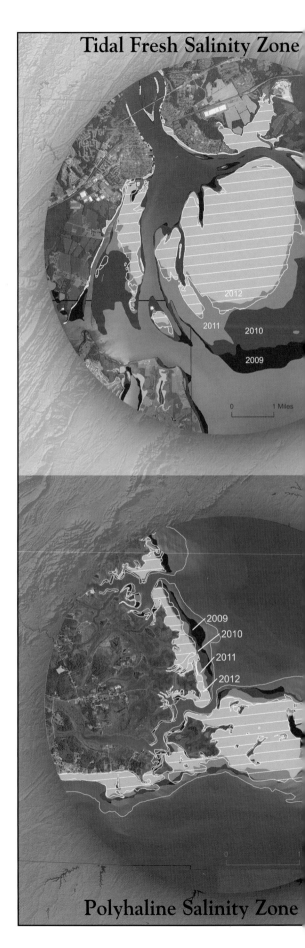

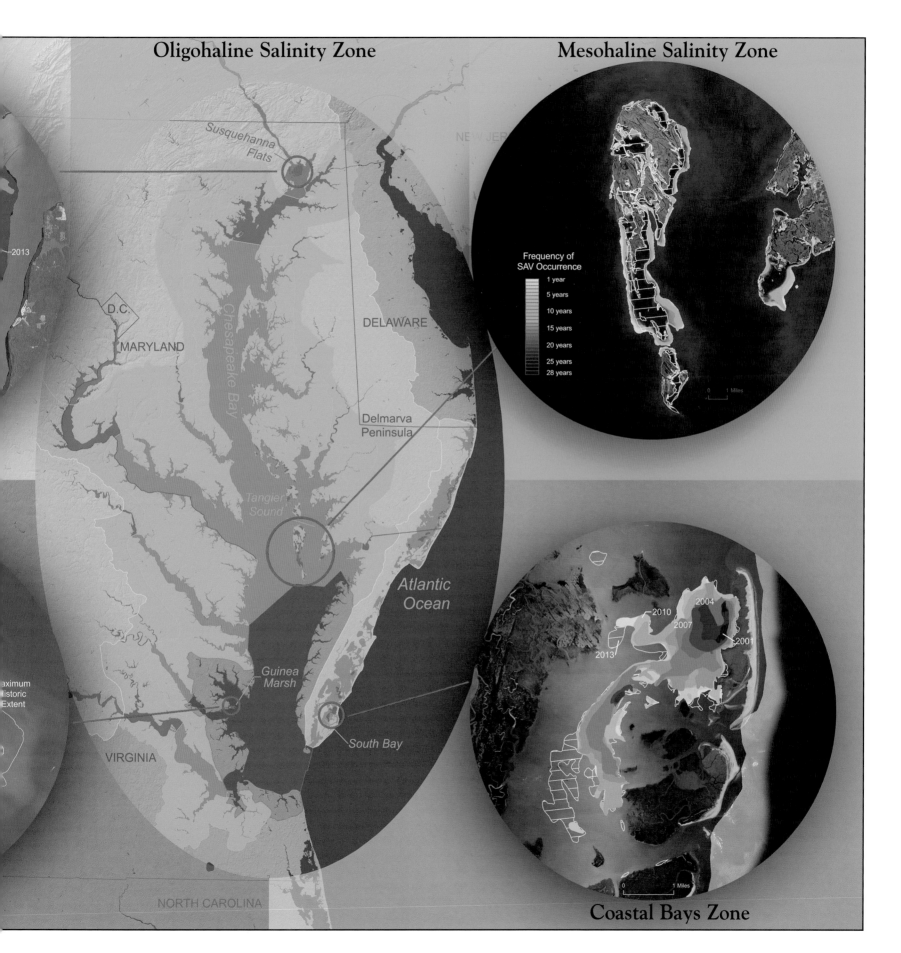

Oligohaline Salinity Zone

Mesohaline Salinity Zone

Coastal Bays Zone

Susquehanna Flats

NEW JERSEY

Chesapeake Bay

D.C.

MARYLAND

DELAWARE

Delmarva Peninsula

Tangier Sound

Atlantic Ocean

Guinea Marsh

South Bay

Maximum Historic Extent

VIRGINIA

NORTH CAROLINA

Frequency of SAV Occurrence

1 year
5 years
10 years
15 years
20 years
25 years
28 years

0   1 Miles

2004
2010
2007
2001
2013

0   1 Miles

Pacific Northwest Marine Ecoregional Assessment Report Maps

The Nature Conservancy (TNC)

Portland, Oregon, USA
By Aaron Jones

Contact
Aaron Jones
ajones@tnc.org

Software
ArcGIS 10.1 for Desktop

Data Sources
College of Earth, Ocean, and Atmospheric Sciences, Oregon State University; Bureau of Ocean Energy Management; Esri; California Natural Diversity Database; BC Ministry of Sustainable Resource Management; National Aeronautics and Space Administration; National Oceanic and Atmospheric Administration; Oregon Department of Land Conservation and Development; Oregon Department of Transportation, Oregon Department of Fish and Wildlife Marine Resources Program; Oregon Biodiversity Information Center; US Census Bureau; US Geological Survey National Gap Analysis Program; US Fish and Wildlife Service; Washington Natural Heritage Program

This map series was part of the Pacific Northwest (PNW) Marine Ecoregional Assessment. The series covered an ecoregion characterized by diverse habitats supporting a broad variety of resident and migratory native species. The PNW was one of over 150 ecoregional assessments completed around the world by The Nature Conservancy (TNC) and its partners over the past twenty years and used TNCs Conservation by Design planning methodology to help prioritize conservation programs. The maps show the ecoregion's protected areas and the assessment units used for evaluating relative biodiversity value and conservation suitability for hundreds of conservation targets over thousands of locations. Maps of the conservation targets themselves include benthic (bottom of a body of water) habitats; species occurrence data estuarine and shoreline habitats, and cold water upwelling and phytoplankton concentrations.

Jones, A. Map Series. 1:100,000. Dec. 2013. In: Vander Schaaf, D., K. Popper, D. Kelly and J. Smith. 2013. Pacific Northwest Marine Ecoregional Assessment. The Nature Conservancy, Portland, Oregon.

Courtesy of The Nature Conservancy.

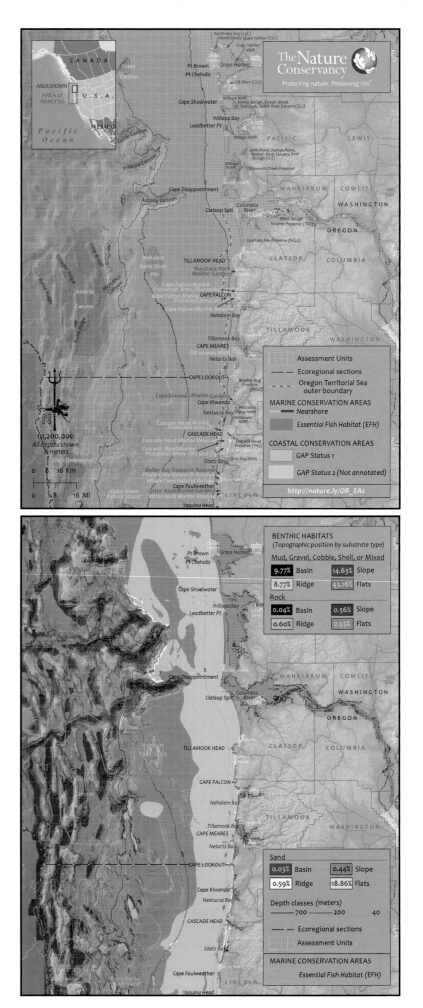

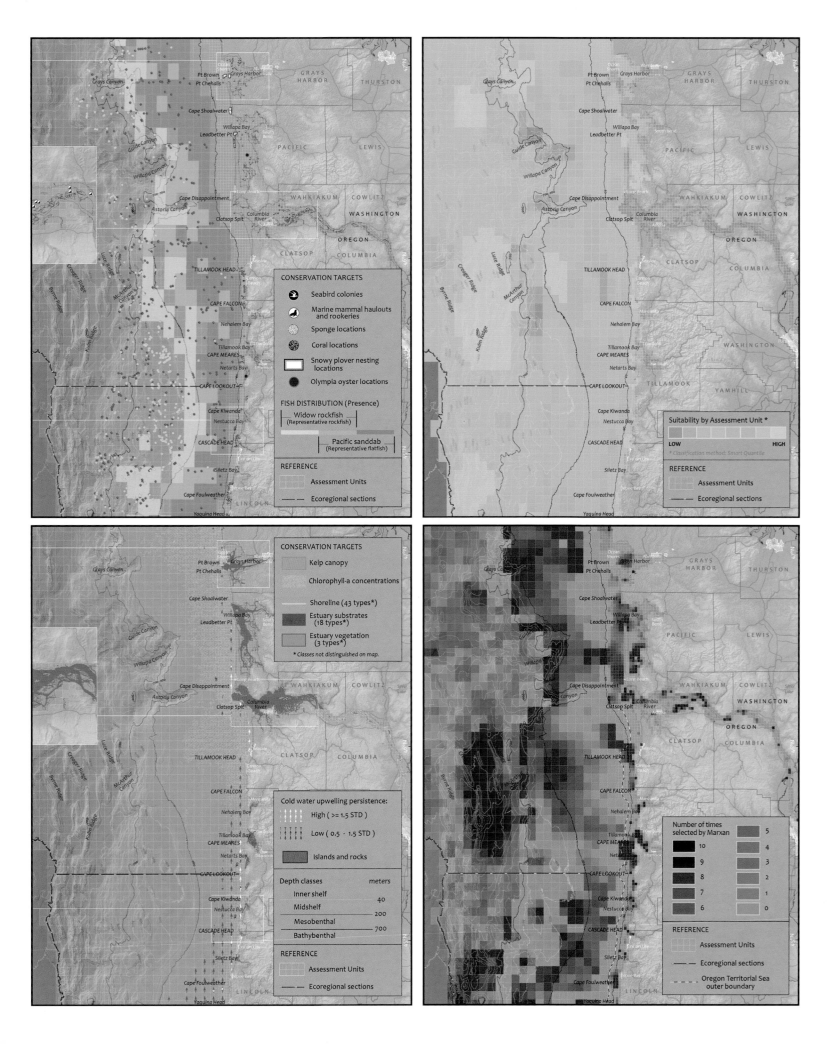

CONSERVATION TARGETS

- Seabird colonies
- Marine mammal haulouts and rookeries
- Sponge locations
- Coral locations
- Snowy plover nesting locations
- Olympia oyster locations

FISH DISTRIBUTION (Presence)

Widow rockfish
(Representative rockfish)

Pacific sanddab
(Representative flatfish)

REFERENCE

Assessment Units

Ecoregional sections

Suitability by Assessment Unit *

LOW   HIGH

* Classification method: Smart Quantile

REFERENCE

Assessment Units

Ecoregional sections

CONSERVATION TARGETS

- Kelp canopy
- Chlorophyll-a concentrations
- Shoreline (43 types*)
- Estuary substrates (18 types*)
- Estuary vegetation (3 types*)

* Classes not distinguished on map.

Cold water upwelling persistence:

High (>= 1.5 STD)

Low (0.5 - 1.5 STD)

Islands and rocks

Depth classes	meters
Inner shelf	
	40
Midshelf	
	200
Mesobenthal	
	700
Bathybenthal	

REFERENCE

Assessment Units

Ecoregional sections

Number of times selected by Marxan

10	5
9	4
8	3
7	2
6	1
	0

REFERENCE

Assessment Units

Ecoregional sections

Oregon Territorial Sea outer boundary

Cell Phone Record Analysis and Visualization

East Bay Regional Park District

Oakland, California, USA
By Kara Hass

Contact
Kara Hass
khass@ebparks.org

Software
ArcGIS 10.1 for Desktop

Data Sources
Cellular company phone records

When a crime is committed on East Bay Regional Park District land, district detectives obtain the suspect's cell phone records and rely on GIS to show the suspect's location before, during, and after a crime. This map shows how to analyze cell phone records (or call detail records) obtained from cellular companies. ArcGIS software is used to illustrate the call location, call frequency, a suspect's movement patterns, and connections between multiple suspects. These maps have been instrumental in the investigation and interrogation of criminals.

Courtesy of East Bay Regional Park District.

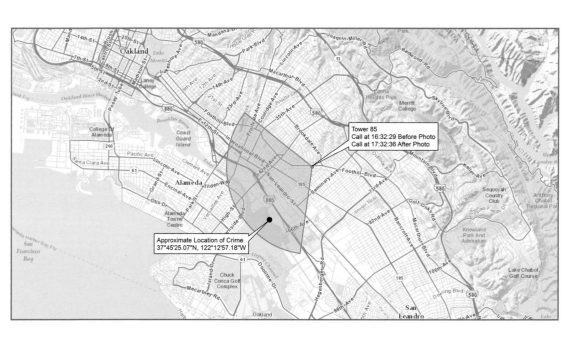

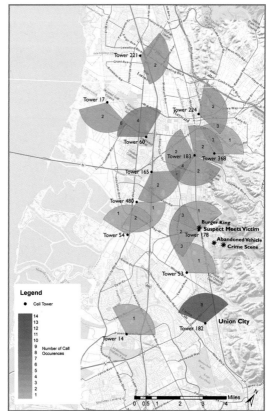

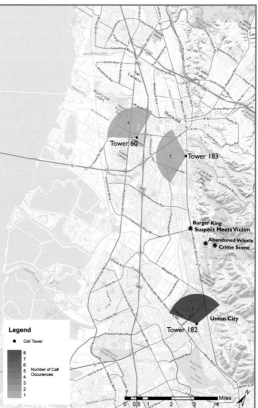

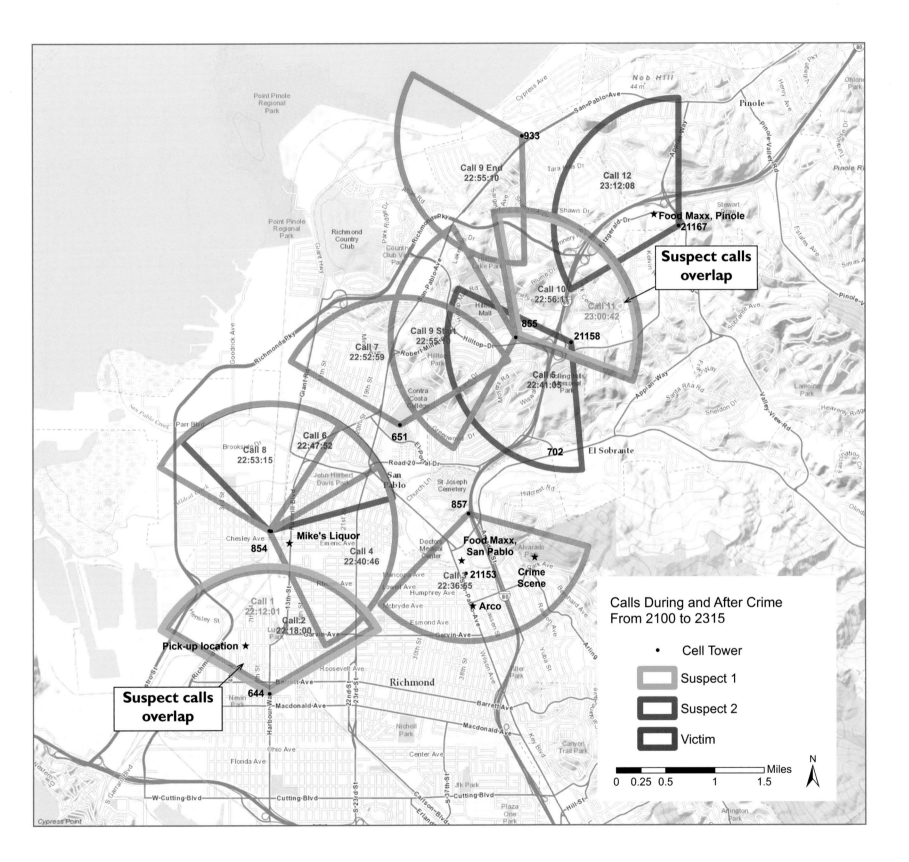

Suspect calls
overlap

Suspect calls
overlap

Calls During and After Crime
From 2100 to 2315

• Cell Tower

Suspect 1

Suspect 2

Victim

Miles
0   0.25 0.5   1   1.5

N

53

Justices of the Peace—Caseload Inputs and Precinct Planning

Collin County

McKinney, Texas, USA

By Bret Fenster, Mohamed Hassan, Kendall Holland, Ramona Luster, Tim Nolan, Gabriela Voicu, and Bill Bilyeu

Contact
Bret Fenster, GISP
bfenster@collincountytx.gov

Software
ArcGIS 10.2 for Desktop

Data Sources
Esri, Collin County, North Central Texas Council of Government, Texas Department of Transportation

Justice of the peace (JP) precincts provide convenient services to citizens in matters of civil and criminal justice. JPs are served by courts located throughout Collin County. The county administrator requested a caseload study to help plan JP redistricting to balance workloads for each precinct. The resulting maps show different densities with the potential inputs to JP caseloads in Collin County. The major inputs are retail locations with 80,000 or more square feet, multifamily housing, traffic counts, toll locations, and population. Collin County's estimated population density for 2013 provides the backdrop for the JP court caseload data. The population density is represented as traditional bar charts and as colored spheres that show the relative civil and criminal workload for each precinct and the corresponding court location. An ArcGIS Online analysis feature service (Create Drive Times) helps display the road networks that surround each court. This study shows that population growth is the driving force behind the increasing caseloads, and the existing precinct configuration provides centralized court locations for the citizens. The data also helps show that redistricting is not always necessary.

Courtesy of Collin County.

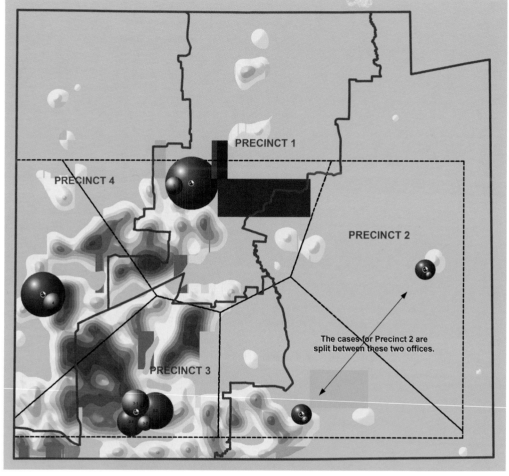

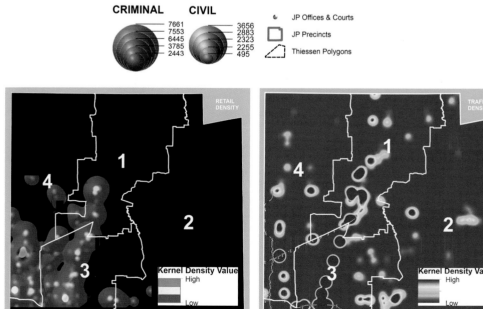

Violence against Women

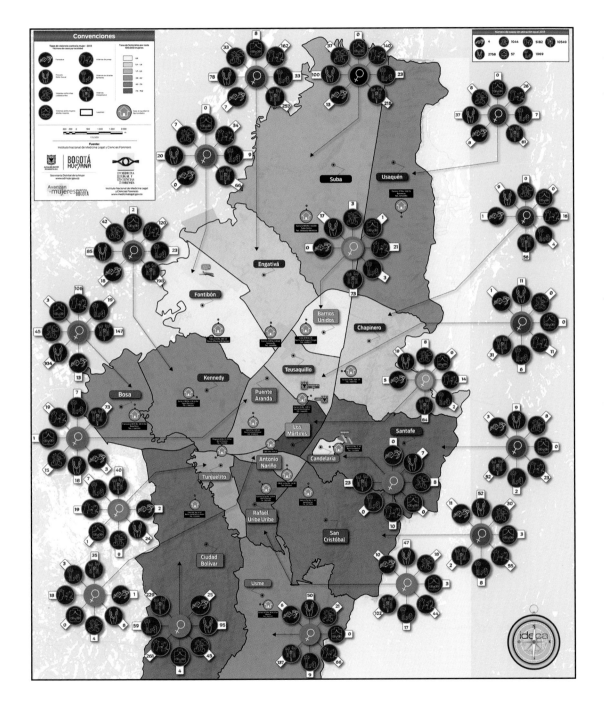

Catastro Distrital (Cadastre District)

Bogota, Cundinamarca, Colombia
By Special Administrative Unit of Cadastre District, Spatial Data Infrastructure for the Capital District

Contact
Carlos Alberto Guarin Ramirez
cguarin@catastrobogota.gov.co
or ideca@catastrobogota.gov.co

Software
ArcGIS 10.1 for Desktop

Data Source
Cadastre Bogotá, Secretary of the Women and National Institute of Legal Medicine

This map was published in Colombia on the occasion of the International Day of Women's Rights. The map locates incidents of violence against women in Bogotá, Colombia's capital, categorized by sexual offenses, violence against children and adolescents, violence against older women, intimate partner violence, violence against other workers, and family and interpersonal violence. In this context, violence against women is a form of discrimination, worsens inequity, inhibits freedom and autonomy of women, and is an insult to human dignity. For this reason, the mayor of Bogotá called on all citizens to recognize the right of women to a life free of violence, to denounce and not legitimize situations of violence, and to punish offenders.

Courtesy of Cadastre District, District Secretariat of Women and National Institute of Legal Medicine.

Land Subsidence Hazard Map of Abandoned Mine Sites in South Korea

Seoul National University

Seoul, Seoul, Republic of South Korea
By Jangwon Suh, Jin Son, Myeongchan Oh,
Hyeong-Dong Park, and Namsoo Choi

Contact
Jangwon Suh
jangwonsuh@hanmail.net

Software
ArcGIS 10.1 for Desktop

Data Source
MGIS of Mine Reclamation Corporation

Subsidence at abandoned mine areas pose serious risks to people, property, and the environment. A Korean mine reclamation company reported that about five thousand mines have been abandoned since the 1990s and many have shown subsidence, especially in residential areas. This study used GIS to map hazards and assess risk in the vicinity of abandoned coal mines. To evaluate mine subsidence risk, the study evaluated factors including drift depth, drift density, distance from nearest drift, distance from nearest railroad, rock mass rating, groundwater depth, slope, and flow accumulation. Researchers mapped subsidence risk by relating subsidence hazard maps and vulnerability maps with residential building maps. The study presents a detailed subsidence risk assessment to support abandoned coal mine area management.

Courtesy of Seoul National University.

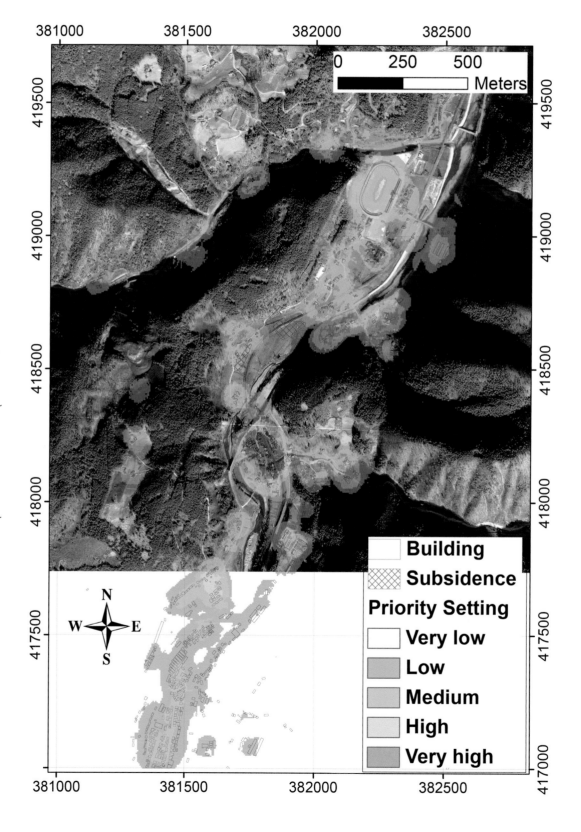

Southern Louisiana Subsidence Vulnerability Estimates

Louisiana State University

Baton Rouge, Louisiana, USA

By Joshua D. Kent, Clifford J. Mugnier, Randy Osborne, Larry Dunaway, and J. Anthony Cavell

Contact
Joshua D. Kent
jkent4@lsu.edu

Software
ArcGIS 10.2.2 for Desktop, Microsoft Excel

Data Sources
Landsat 5 Thematic Mapper, National
Elevation Dataset, National Geodetic
Survey Technical Report 50

Accurate, reliable, and consistent elevation information contributes to informed decisions for a wide range of important activities, including mapping and charting, determining flood risk, land-use planning and management, environmental management, economic development, coastal sustainability, and engineering design. In no place is this more critical than Louisiana. In addition to sea level rise, Louisiana is facing unprecedented rates of land loss—up to a football field every hour—contributing to an increased risk from flooding hazards. A report recently published by the National Research Council projects significant annual flood losses for New Orleans. In fact, New Orleans ranks fourth in the world for flood loss. This map reveals the extent and the consequences of long-term subsidence. The map also shows that if trends continue unmitigated, coastal subsidence and land loss will threaten sustainability in Louisiana.

Courtesy of Louisiana State University.

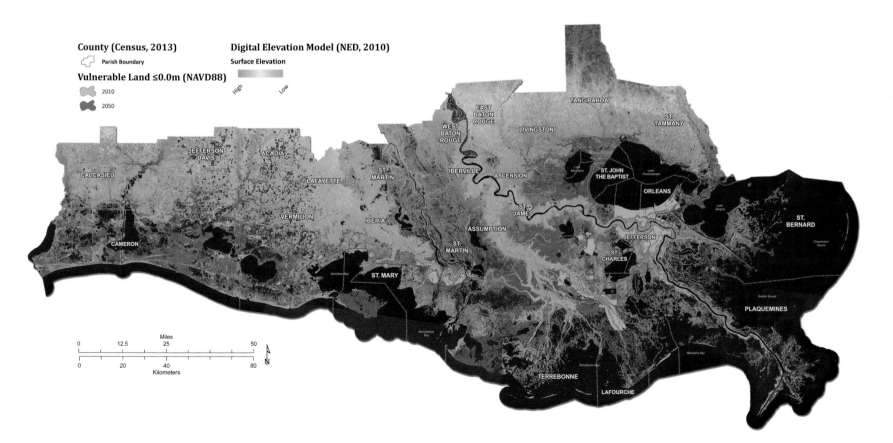

Map of Fire Risk Area for Managing Disaster

Daewon Aero Survey Co., Ltd.

Seocho-gu, Seoul, Republic of South Korea
By Wan-Yeong Song, Jong-Bae Kim, and Kwang-Hyun Cho

Contact
Wan-Yeong Song
it4korea@naver.com

Software
ArcGIS 10.1 for Desktop, Adobe Photoshop

Data Source
Firefighting field data and national digital topographic map

This map shows the fire risk areas of Daegu, the third largest metropolitan area in South Korea with over 2.5 million residents. Daegu is located in southeastern Korea about 50 miles from the seacoast. The map shows where risk of fire property loss and injury is likely to be greatest due to reduced firefighting capability. The firefighting activities were defined in three stages: rushing to the site, operation of resources, and fighting the fire. The critical factors were defined for each firefighting activity to prepare the risk map

Courtesy of Daewon Aero Survey Co., Ltd.

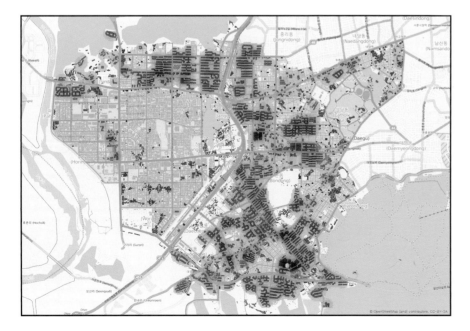

Legend

The width of road is not more than 4m

Fire engine cannot access due the parked vehicles

Road

The area where the fire engine can access

The area where the fire engine cannot access

Fire risk area was prepared using the distance between hydrants and the target

The network analysis was conducted using the position of fire fighting water

Object-specific fire

First grade
Second gra
Special gra

DAEGU in Republic of Korea

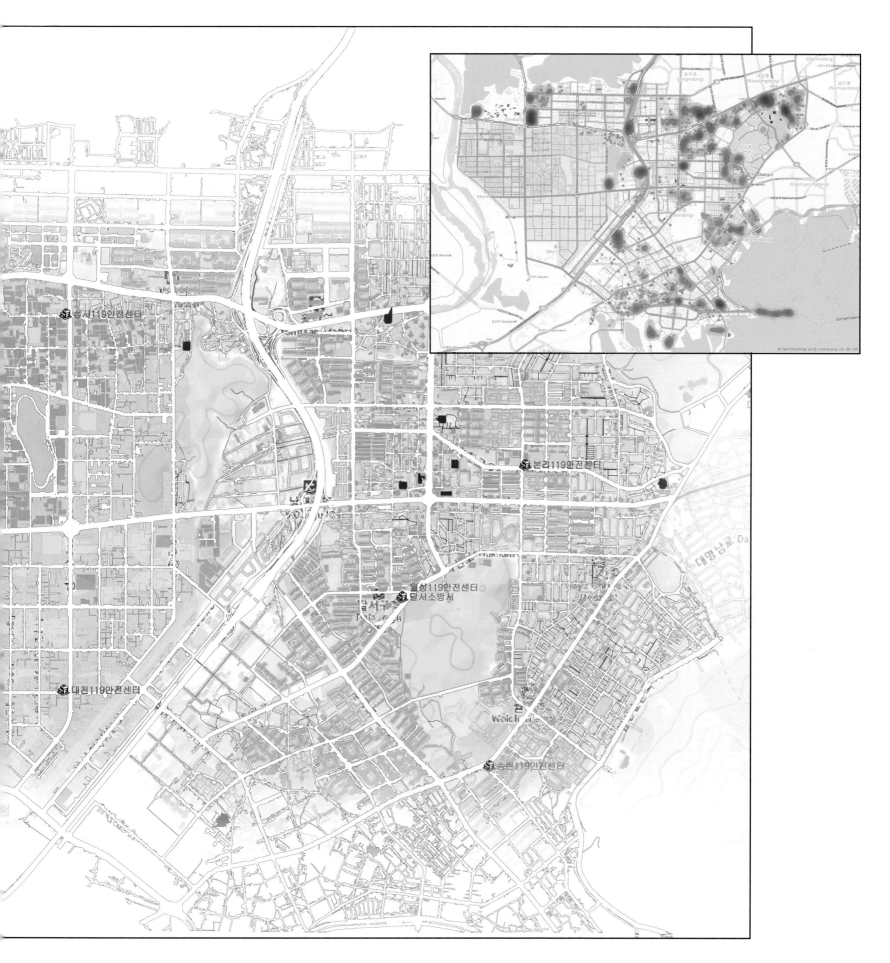

성서119안전센터

본리119안전센터

월성119안전센터
달서소방서

대천119안전센터

송현119안전센터

59

All Hazards Map

City of Chula Vista

Chula Vista, California, USA
By Rommel Reyes

Contact
Rommel Reyes
rreyes@chulavistaca.gov

Software
ArcGIS 10.2.1 for Desktop

Data Sources
City of Chula Vista, California Emergency Management Agency, California Geological Survey, California Department of Forestry and Fire Protection, Federal Emergency Management Agency, US Geological Survey

As part of an effective emergency response plan, the City of Chula Vista Emergency Operations Center (EOC) created a citywide map identifying all critical facilities and the potential hazards that could affect them. Working with key members of the Development Services and Public Safety departments, the GIS staff designated over 300 facilities as critical during a significant emergency event. In addition, the members identified potential hazards that could pose a significant threat to the residents of Chula Vista that include flooding, earthquakes, and wildfires. With the completed map, EOC staff members now have access to critical baseline data to aid in developing an effective emergency response. Depending on the type, size, and location of the hazard, the EOC can quickly identify and alert critical assets at risk, position public safety resources to high-priority areas, establish evacuation routes, and designate certain critical facilities as emergency shelters.

Courtesy of City of Chula Vista.

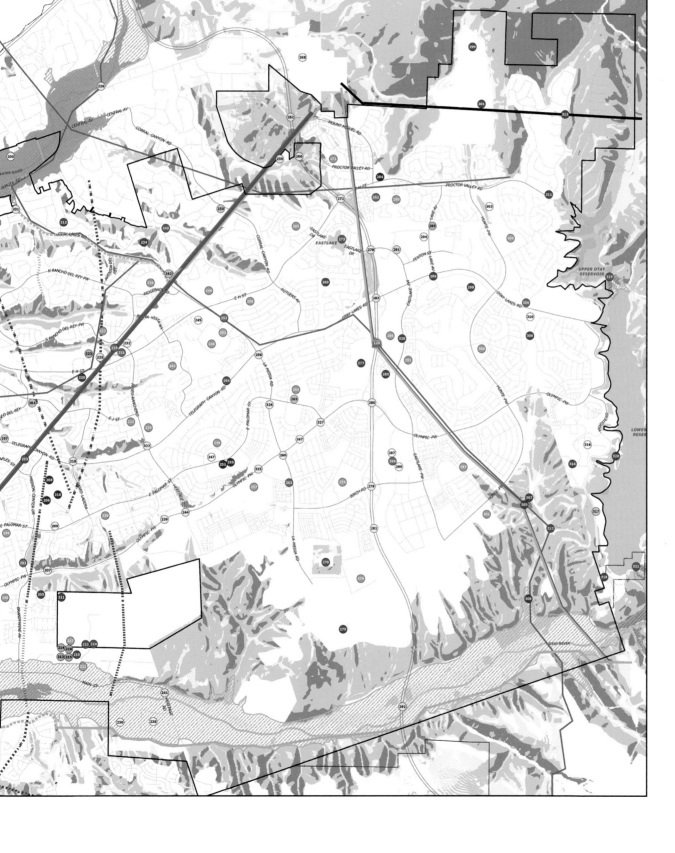

Potential Hazards

┈┈┈┈ Faults

▨ Inundation

█ 100 ft above sea level

█ 40 ft above sea level

Fire Hazard Rating

Low

Moderate

High

Very High

✖✖✖✖ Tsunami Evacuation Line

Critical Facilities

○ Bridges

● Business Facilities

● Communication Facilities

● Dams

○ Electric Power Facility

● Emergency Centers/Fire/Police

○ Govt Offices/Civic Centers

● Hospitals/Care Facilities

● Jails/Prisons

○ Marinas

● Natural Gas

○ Other Schools

○ Port Facilities

○ Post Offices

● Potable/Waste Water Facilities

● Schools

○ Tourist Attractions

● Transit Facilities

—— 069kV, OH

┈┈┈ 069kV, UG

—— 138kV, OH

┈┈┈ 138kV, UG

—— 230kV, OH

┈┈┈ 230kV, UG

—— 500kV, OH

—— Natural Gas

▨ Waste Staging Site

VICINITY MAP

Municipal Asset Management in Golden, BC

Town of Golden

Golden, British Columbia, Canada
By Alyson Marjerrison

Contact
Alyson Marjerrison
alyson.marjerrison@golden.ca

Software
ArcGIS for Desktop 10.2

Data Source
Town of Golden

Municipalities use asset management to keep an accurate record of the worth, projected lifespan, and current condition of their infrastructure. Municipalities must plan for inevitable renewal of their infrastructure. The Town of Golden took three components of its linear assets (waterlines, sewer lines, and roads) and rated each condition independently. Once the town calculated the utility conditions, it identified hot spots that had the worst rating. Next, the town gave nineteen projects priority for infrastructure renewal over a five- or ten-year time frame. Proposed infrastructure renewal projects were presented to the town's council in a mapped infographic that outlined details such as asset condition, age, and cost. This strategic initiative put Golden in a prime position ahead of other municipalities when applying for Building Canada government grants.

Courtesy of Alyson Marjerrison, Town of Golden.

Jefferson County, Colorado, Parcel Map Series

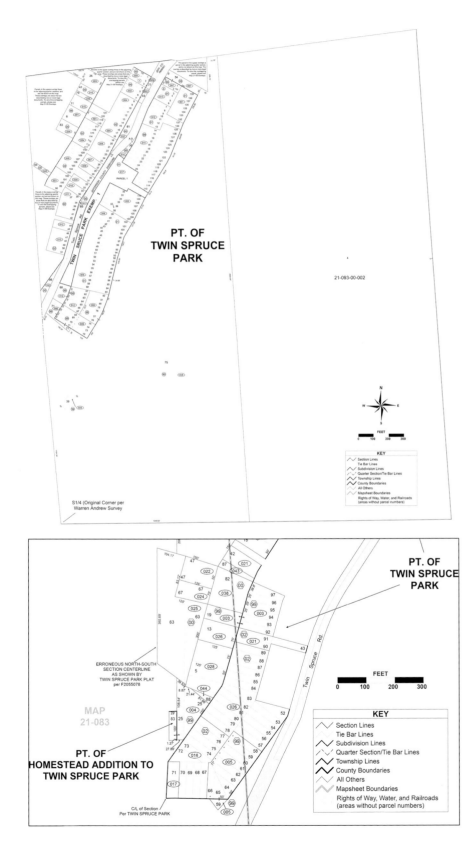

Jefferson County Assessor's Office

Golden, Colorado, USA
By John N. Hansen, GISP

Contact
John N. Hansen
jnhansen@jeffco.us

Software
ArcGIS 10.1 for Desktop, Sidwell Parcel Builder MapPlotter 4.1 SP 1

Data Sources
Jefferson County cadastral geodatabase, Microsoft SQL Server 2008 R2

The Jefferson County Assessor's Office maintains 1,440 parcel maps which cover approximately 780 square miles. The county contains over 215,000 parcels. This subset of production parcel maps covers an area of overlapping parcels within a mountainous and rugged region in northwestern Jefferson County. The maps were designed to clearly and impartially show each of the affected parcels. Numerous contributing factors have led to these parcel overlaps. When townships were subdivided into sections by the Government Land Office (GLO, now the Bureau of Land Management) in 1870, the center quarter section corner of section 8 was not set, which was the standard operating procedure. A proper breakdown of section 8 was not done until 1974. Neither the Twin Spruce Park Subdivision Plat from 1922, nor the Homestead Addition to Twin Spruce Park Plat from 1927, were properly tied to section or quarter-section corners. Additionally, property pins were probably not set at the time of platting. Deeds and land survey plats call out an "erroneous north-south centerline of Section," and four possible north-south centerlines for section 8 are shown on a recorded Improvement Survey Plat and Parcel Exhibit. And finally, metes and bounds deeds for parcels near the subdivisions overlap each other.

Thanks to Michael M. Greer, cartographic supervisor with Jefferson County Planning and Zoning, for help with this research.

Courtesy of Jefferson County Assessor's Office.

Developable Land in Central Arizona

Maricopa Association of Governments (MAG)

Phoenix, Arizona, USA
By Kurt Cotner and Jason Howard

Contact
Jason Howard
jhoward@azmag.gov

Software
ArcGIS 10.2.1 for Desktop

Data Sources
Maricopa Association of Governments, Central Arizona Governments, Arizona State Land Development

This map depicts developable land in Maricopa and Pinal Counties in Arizona. The developable land dataset gives the Maricopa Association of Governments' member agencies a clearer picture of the availability of land for growth. In particular, this data points to the importance of State Trust land in that growth. These lands were granted to the state under the federal act that provided for Arizona's statehood in 1912. Within the two-county region, approximately 29 percent of the land is classified as developable, with State Trust land accounting for 62 percent of that. Large blocks of State Trust land on the urban periphery are both an opportunity and a constraint to future development. While these large tracts of land offer ample opportunity for the development of master planned communities, solar generating stations, and open space, the process of developing these tracts is slowed by the auction process due to state, public, and political scrutiny of that process, and lack of infrastructure on the lands in question.

Courtesy of Maricopa Association of Governments and the MAG member agencies.

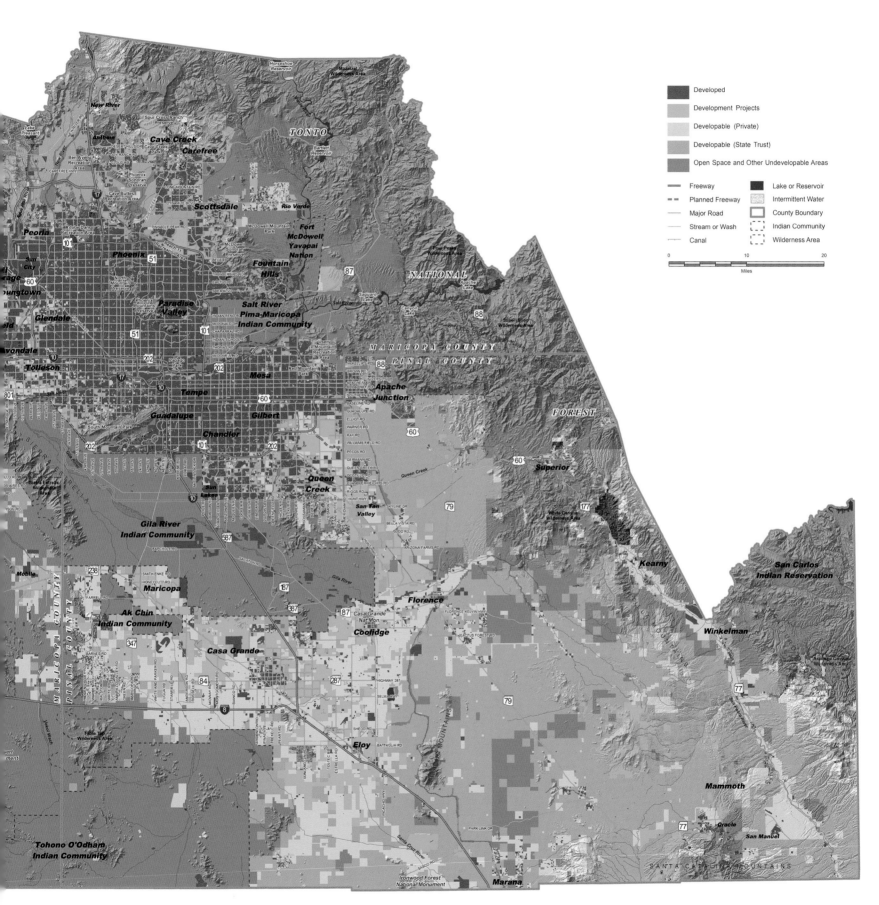

Developed

Development Projects

Developable (Private)

Developable (State Trust)

Open Space and Other Undevelopable Areas

Freeway   Lake or Reservoir
Planned Freeway   Intermittent Water
Major Road   County Boundary
Stream or Wash   Indian Community
Canal   Wilderness Area

0   10   20
              Miles

65

Estados Unidos Mexicanos (United States of Mexico)

Cobb, Fendley & Associates, Inc.,

Houston, Texas, USA
By Larry Jahn

Contact
Larry Jahn
ljahn@cobbfendley.com

Software
ArcGIS 10.2 for Desktop

Data Sources
Esri, Instituto Nacional de Estadistica y Geografia, DeLorme Publishing Co. Inc., GADM.org, and Digital Chart of the World

This is a hybrid political map by civil engineering firm Cobb, Fendley & Associates that blends traditional cartographic techniques with the topographical beauty of the Esri ocean basemap. The map shows the boundaries of Mexico's thirty-one federal states and the Mexico City Federal District, transportation networks, hydrography, and points of interest.

Courtesy of Cobb, Fendley & Associates, Inc.

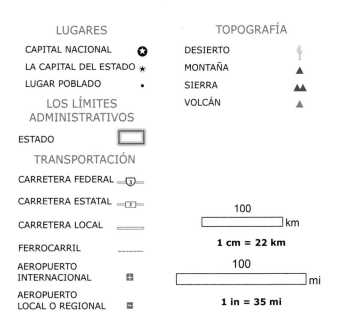

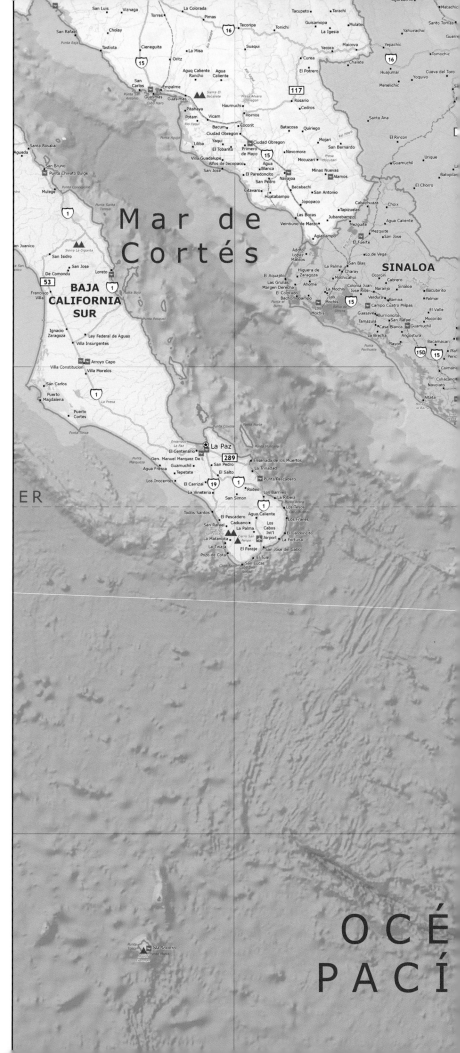

Town of Mattapoisett Assessment Neighborhood Maps

Applied Geographics, Inc.

Boston, Massachusetts, USA
By Michele Giorgianni, Danielle Gwinn, Peter Lemack, and David Weaver

Contact
Richard Grady
rgrady@appgeo.com

Software
ArcGIS 10.1 for Desktop

Data Sources
Town of Mattapoisett Computer-Assisted Mass Appraisal database, Town of Mattapoisett cadastral dataset

The proximity of properties to amenities affects property valuation in ways that assessors need to account for. Municipalities with extensive waterfronts, such as the Massachusetts seaside town of Mattapoisett, must balance neighborhood and individual property valuations. These maps facilitate review and quality control by the assessor. The "Neighborhood Location Factor Codes" map presents the results of a systematic method for valuing properties based on proximity to desirable locations (waterfront) or amenities (golf course), and other desirable neighborhood-level characteristics. The map combines groups of parcels assigned the same valuation factor for further analysis and use by the assessor. A complementary Special Parcel Location Property Codes Map (not shown) displays parcel-level coding based on locational characteristics that also help the assessor resolve anomalies between neighborhood boundaries and individual property land values.

Courtesy of Town of Mattapoisett and Applied Geographics, Inc.

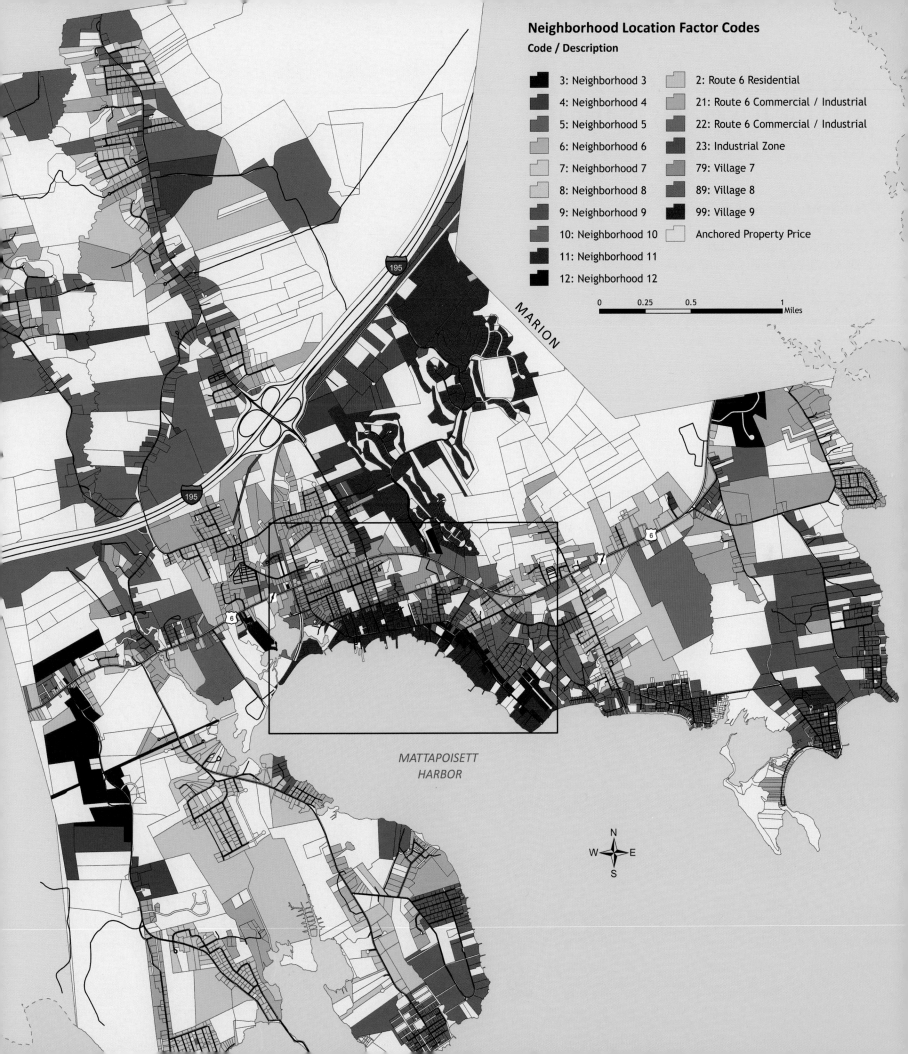

Neighborhood Location Factor Codes

Code / Description

- 3: Neighborhood 3
- 4: Neighborhood 4
- 5: Neighborhood 5
- 6: Neighborhood 6
- 7: Neighborhood 7
- 8: Neighborhood 8
- 9: Neighborhood 9
- 10: Neighborhood 10
- 11: Neighborhood 11
- 12: Neighborhood 12
- 2: Route 6 Residential
- 21: Route 6 Commercial / Industrial
- 22: Route 6 Commercial / Industrial
- 23: Industrial Zone
- 79: Village 7
- 89: Village 8
- 99: Village 9
- Anchored Property Price

0   0.25   0.5   1
Miles

MARION

MATTAPOISETT HARBOR

N
W   E
S

Shaded Topography of Indiana

Indiana Geological Survey

Bloomington, Indiana, USA
By Matthew R. Johnson and Kevin P. Russell

Contact
Matthew R. Johnson
mrj21@indiana.edu

Software
ArcGIS 10.0 for Desktop, Microsoft Office, Adobe Creative Suite 6

Data Sources
IndianaMap, Indiana Spatial Data Portal, OpenStreetMap

This map provides a highly detailed depiction of the diverse landscape of Indiana and showcases the most recent lidar elevation data for the entire state acquired over three years. The map supports decision making where topographic features are of critical importance. The map also facilitates conversations among educators, legislators, and stakeholders regarding the importance of timely and high-quality data acquisition. The high-resolution digital elevation model (DEM) provides an elevation value for every 5-square-foot pixel shown on the map. The Indiana Geological Survey serves the entire DEM as a mosaic dataset for internal geologic research and as cached tiles for online viewing through the IndianaMap website, the largest publicly available collection of Indiana GIS map data.

Courtesy of Indiana Geological Survey.

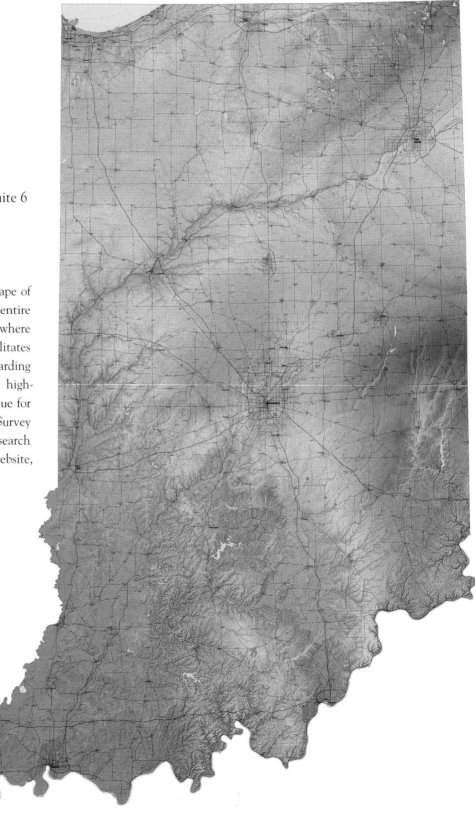

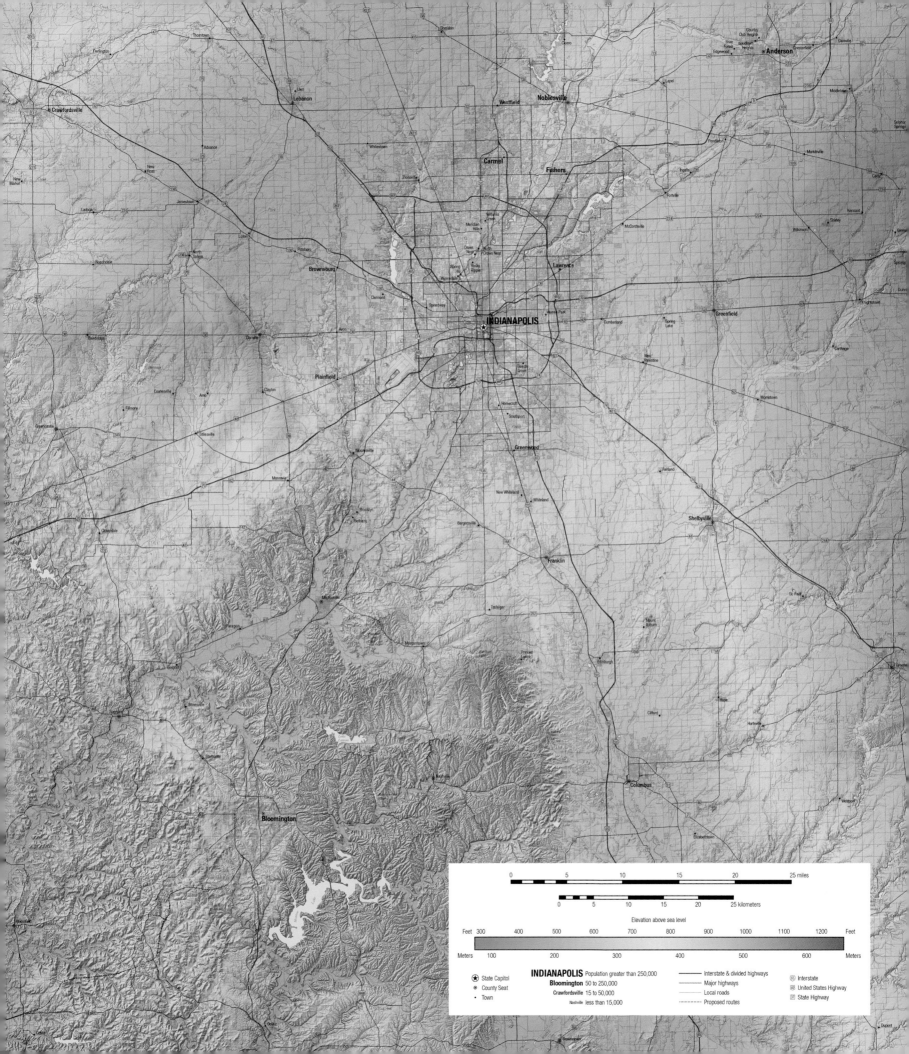

INDIANAPOLIS

Population greater than 250,000

Bloomington 50 to 250,000

Crawfordsville 15 to 50,000

Nashville less than 15,000

⭐ State Capitol

⊛ County Seat

• Town

Elevation above sea level

| Feet | 300 | 400 | 500 | 600 | 700 | 800 | 900 | 1000 | 1100 | 1200 | Feet |
| Meters | | 100 | 200 | | 300 | | 400 | | 500 | 600 | Meters |

Interstate & divided highways

Major highways

Local roads

Proposed routes

Interstate

United States Highway

State Highway

The Demographics of Zoning

Portland State University

Portland, Oregon, USA
By Richard Lycan

Contact
Richard Lycan
lycand@pdx.edu

Software
ArcGIS 10.2.1 for Desktop

Data Sources
2010 US Census, National Center for Health Statistics, Oregon Geospatial Enterprise Office

The Oregon legislature directed the Population Research Center at Portland State University to develop coordinated population forecasts for the state's counties and cities. Oregon land use planning law requires all cities and counties to develop zoning plans. Zoning maps may provide guidance for the population forecasts. State officials needed, in particular, to determine where the rapidly increasing numbers of older persons will reside. The map "Where the Seniors Live" for the Portland Metro area shows various contexts in which seniors are found. In yellow, 20 to 49 percent of the population is age 55 and over, most living in single family housing and attempting to age in place. In green, 0 to 19 percent of the population is age 55 and over, but many less affluent older persons reside in this zone. In red and purple, over 50 percent of persons age 55 plus live in senior housing and commercial apartments. This block level census data also was used for the demographic analysis. The map "Generalized Zoning" is from statewide coverage from the Oregon Geospatial Enterprise Office. While a statewide standard governs the zoning classes, the application varies from county to county. The effect of the urban growth boundary can be seen in the abrupt change from urban to rural zoning classes at the growth boundary in the "Generalized Zoning" map. The net migration diagram shows loss of older people through death and net out-migration as well as loss of people in their twenties though out-migration. The diagram also shows net in-migration of older households and their children.

Courtesy of Portland State University.

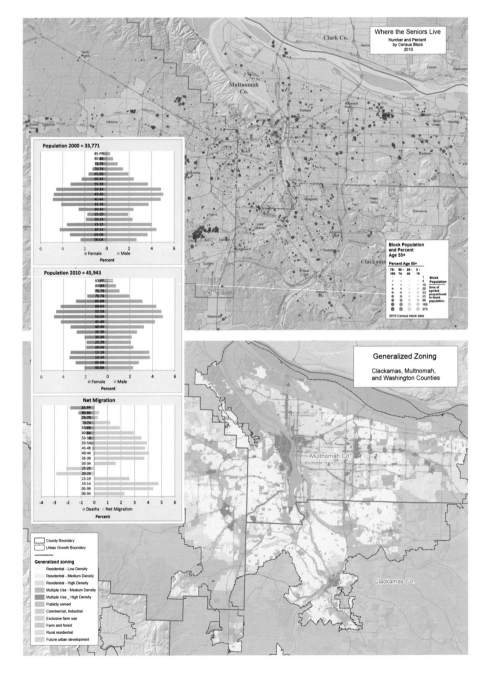

Land-Use Distribution on Abu Dhabi Island

Abu Dhabi City Municipality

Abu Dhabi, Abu Dhabi, United Arab Emirates
By Giridhar Reddy Kolan

Contact
GIS Support
GIS.Support@adm.abudhabi.ae

Software
ArcGIS 10.2.1 for Desktop

Data Source
Abu Dhabi City Municipality

Land-use practices vary across Abu Dhabi Island. Land-use codes can be used to evaluate and assess land disposition and use, which helps decision makers plan better cities. The Abu Dhabi City Municipality's plot database was used to create this land use distribution map. In this map only commercial, educational, governmental, religious, residential, and utility classes have been used to show their distribution per sector on Abu Dhabi Island.

Courtesy of Abu Dhabi City Municipality, Spatial Data Division.

Greensboro Building Permit Activity

City of Greensboro

Greensboro, North Carolina, USA
By Todd Hayes

Contact

Todd Hayes
todd.hayes@greensboro-nc.gov

Software

ArcGIS 10.1 for Desktop

Data Source

City of Greensboro building permit data

Building permits are one type of economic indicator for the City of Greensboro. This map depicts indicators and metrics associated with building permit activity within the city for fiscal year 2013–2014. The main map displays not only the location of every permit issued for the year but also a quantitative graduated dollar value for that particular project. Other maps depict the number of permits by council district, census block, and density of permits. The building inspections database contains rich, multivariate data, such as structure type or occupancy type, that can be displayed spatially or in a tabular format.

Courtesy of City of Greensboro Development Services, Greensboro GIS.

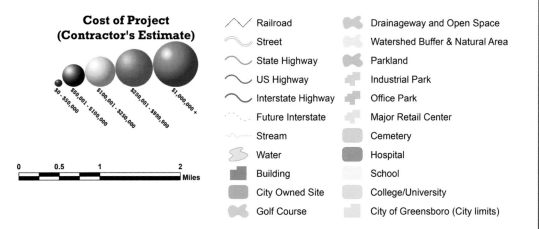

Cost of Project (Contractor's Estimate)

$0 - $50,000 $50,001 - $100,000 $100,001 - $250,000 $250,001 - $999,999 $1,000,000 +

Railroad	Drainageway and Open Space
Street	Watershed Buffer & Natural Area
State Highway	Parkland
US Highway	Industrial Park
Interstate Highway	Office Park
Future Interstate	Major Retail Center
Stream	Cemetery
Water	Hospital
Building	School
City Owned Site	College/University
Golf Course	City of Greensboro (City limits)

0   0.5   1   2 Miles

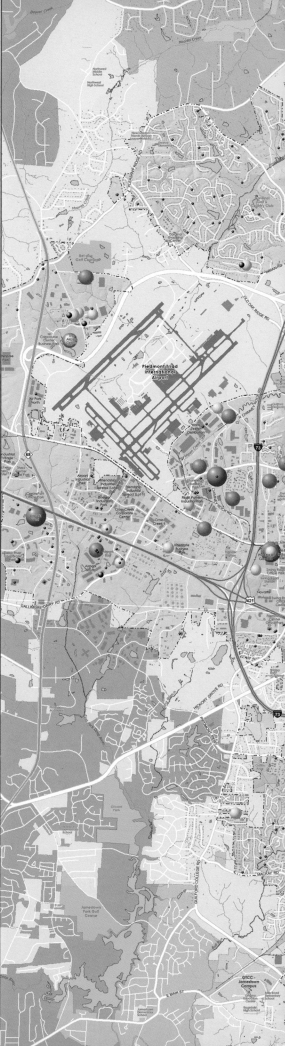

Improving Service Delivery through the Tapestry of Geodesign

Department for Communities and Social Inclusion

Adelaide, South Australia, Australia
By Gary Maguire and David Coombe

Contact
Gary Maguire
gary.maguire@sa.gov.au

Software
ArcGIS 10.2.1 for Desktop

Data Source
South Australian Government

The city of Adelaide is designed with a grid-type layout consisting of branches ending in cul-de-sacs. The city is also overlaid by a "spoke and wheel" public transport system and the spatial distribution of social housing. Adelaide's geodesign, which couples physical design with systems and behaviors, lets the Department for Communities and Social Inclusion ask what-if questions. The evidence gathered from these questions helps decision makers improve service delivery in line with the department's mission: A better life for South Australians.

Courtesy of Department for Communities and Social Inclusion.

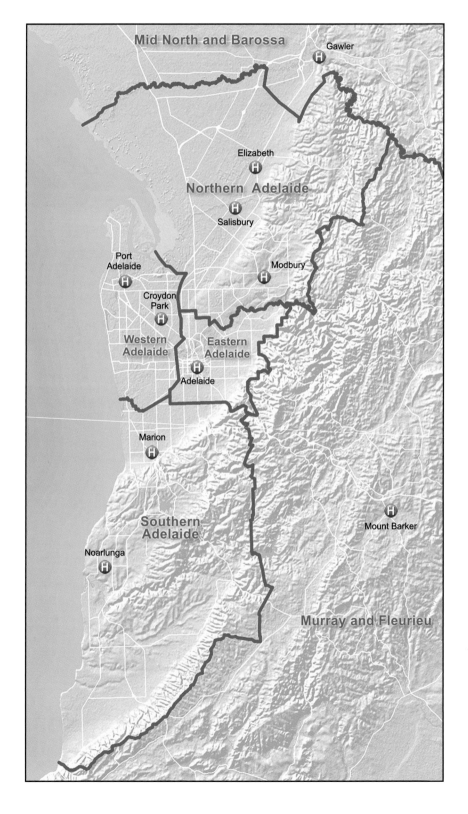

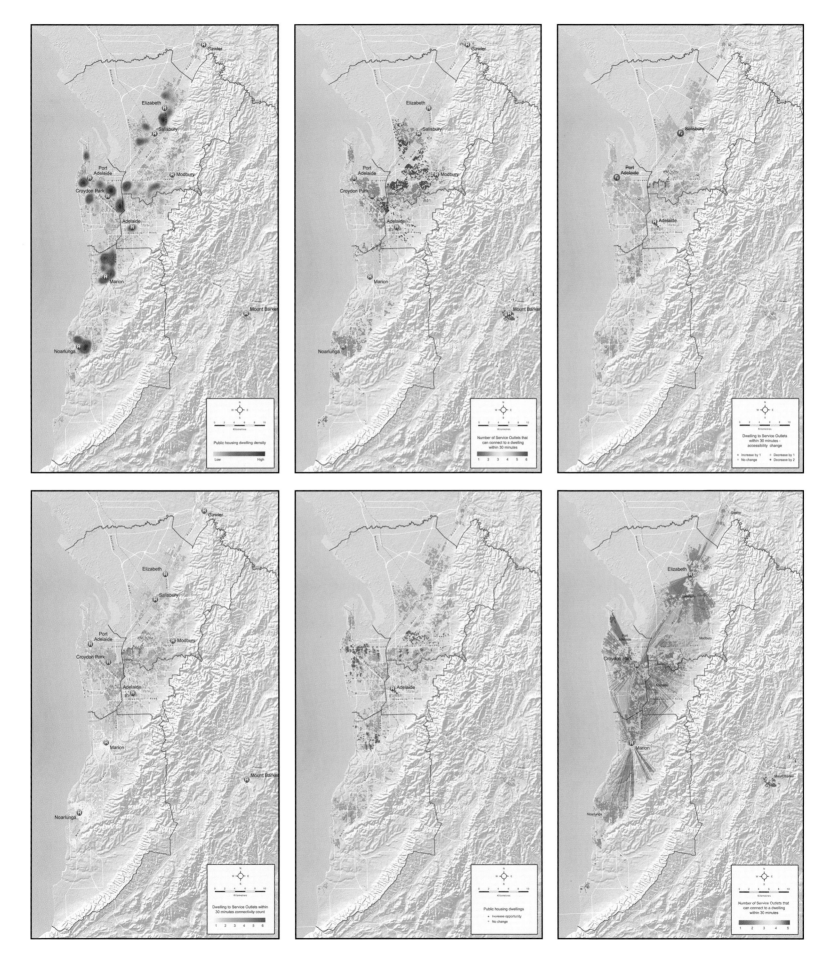

City of Oxnard 2030 Land-Use Map

City of Oxnard

Oxnard, California, USA
By Salvador Mancha

Contact
David Endelman
david.endelman@ci.oxnard.ca.us

Software
ArcGIS 10.1 for Desktop, Adobe Illustrator CS5

Data Source
City of Oxnard

The City of Oxnard 2030 General Plan sets out a vision to guide future development in the California coastal city to the year 2030. The *2030 Land-Use Map* classifies and displays envisioned community land uses and intensity. The map depicts the City Urban Restriction Boundary established by the 1998 Save Open Space and Agricultural Resources ordinance. Planning maps like this used with emerging geodesign concepts and practices will help the City of Oxnard prepare for its future.

Courtesy of Salvador Mancha, City of Oxnard.

Residential

RL	Low
RLM	Low Medium
RM	Medium
RMH	Medium High
RH	High
MHP	Mobile Home Park

Commercial

CCV	Convenience
CN	Neighborhood
CCM	Community
CG	General
CR	Regional
COF	Office
CBD	Central Business District

Industrial

ILM	Limited
ILT	Light
IH	Heavy
CIA	Central Industrial Area
BRP	Business Research Park

Open Space/Other

PRK	Park
RP	Resource Protection
OS	Open Space
PR	Planning Reserve
ESM	Easement
AC	Airport Compatible
AG	Agriculture (County Of Ventura)
SCH	School
PSP	Public/Semi Public
PUE	Energy Facility
❋	Urban Village

Coastal Zone Areas

REX	Residential Existing
RHD	Residential High Density
PUD	Planned Unit Development Residential
✳	Mixed Use
MHP	Mobile Home Park Coastal
HCI	Harbor Channel Islands
VSC	Visitor Serving Commercial
REC	Recreation Area
RP	Resource Protection
PUB	Public Facility
ICD	Industry Priority To Coastal Dependent
PUE	Public Utility/Energy Facility

GIS in Health Impact Assessments

CSS-Dynamac and US Environmental Protection Agency (EPA)

Cincinnati, Ohio, USA

By Amy Prues, Alexander Hall, Florence A. Fulk, Lauren Adkins, and Justicia Rhodus

Contact

Amy Prues

prues.amy@epa.gov

Software

ArcGIS 10.2.1 for Desktop

Data Sources

US Census Bureau, National Oceanic and Atmospheric Administration National Climatic Data Center, City of Atlanta, Fulton County Board of Assessors, US Geological Survey, Landsat Thermal Remote Sensing Toolbox

These maps highlight the geospatial analysis and data the US Environmental Protection Agency used in two health impact assessments of the Proctor Creek watershed near Atlanta, Georgia. Shown here are techniques for identifying vulnerable populations, evaluating the effectiveness of green infrastructure installations, examining the impacts of extreme heat events, and locating deteriorated properties. Demographic information was compiled by tract to identify populations most vulnerable to environmental and social stressors and communities that would benefit most from proposed improvements. Land surface temperature (LST) is an effective way of estimating the impact of hot days that can lead to heat-related illnesses and death. Parcels without central air conditioning were also mapped to identify areas where populations face greater risk from extreme heat events. Deteriorated properties may lead to social and economic decline. Areas with a concentration of such properties would benefit from proposed improvements.

Courtesy of CSS-Dynamac and US Environmental Protection Agency.

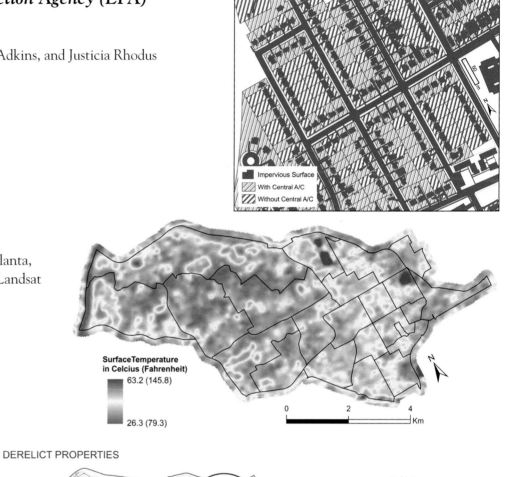

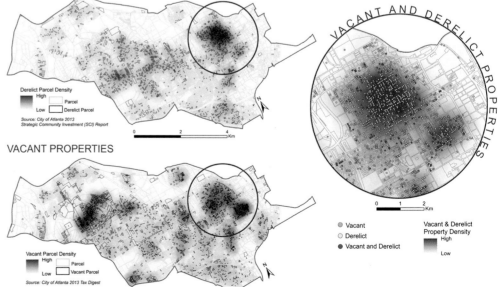

Stage of Breast Cancer and Distance between Patient and Closest Facility

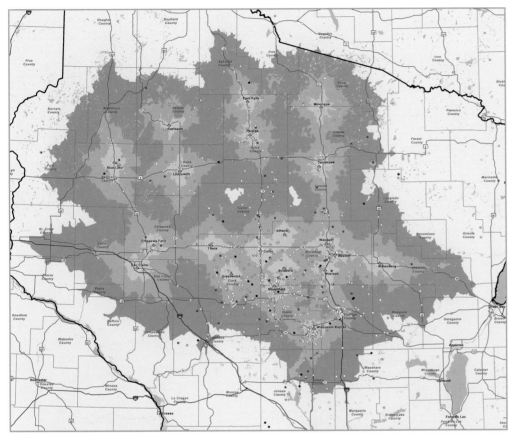

University of Wisconsin-Stevens Point

Stevens Point, Wisconsin, USA
By Michael Broton and Douglas Miskowiak

Contact
Douglas Miskowiak
doug.miskowiak@uwsp.edu

Software
ArcGIS 10.0 for Desktop, ArcGIS Business Analyst, Adobe Illustrator

Data Sources
Marshfield Clinic Cancer Registry, Esri StreetMap Premium for ArcGIS with TeleAtlas 2010

This map shows the relationship between the stage of breast cancer in female patients and the distance between their home and the closest mammogram facility. In the studied area, travel time to the nearest mammogram center appears inversely related to regular mammography screening and breast cancer stage at diagnosis. Women with no missed mammograms before diagnosis lived a median of 15 minutes from the nearest facility. Those who missed five of their past five mammograms lived nearly twice as far, with a median travel time of 27 minutes. The study found a direct relationship between travel time to the nearest mammogram facility and the stage of breast cancer at diagnosis. Travel time increased from 17 to 24 minutes for stage 0 and stage 4 breast cancers, respectively.

Courtesy of Michael Broton and Douglas Miskowiak, University of Wisconsin-Stevens Point; Adedayo Onitilo, Jessica Engel, Hong Liang, and Rachel Stankowski, Marshfield Clinic; and Suhail Doi, University of Queensland.

- Stage 0 Breast Cancer (303 Patients)
- Stage 1 Breast Cancer (603 Patients)
- Stage 2 Breast Cancer (304 Patients)
- Stage 3 Breast Cancer (112 Patients)
- Stage 4 Breast Cancer (52 Patients)
- Stage Unknown Breast Cancer (47 Patients)
- Ministry & Marshfield Clinic Facilities (22 Facilities)
- 0 to 5 Minutes Service Area Buffer Zones
- 5 to 15 Minutes Service Area Buffer Zones
- 15 to 30 Minutes Service Area Buffer Zones
- 30 to 60 Minutes Service Area Buffer Zones
- Outside 60 Minutes Service Area Buffer Zones
- Water Body
- Interstate Highway
- US and State Highways
- State Border
- County Border

0 to 5 Minute Service Area Buffer Zones		
Stage of Breast Cancer	Number of Patients	Percent of Patients
0	54	20.38
1	119	44.90
2	57	21.51
3	22	8.30
4	8	3.02
Unknown	5	1.89
Total Patients 5 Mins	265	100.00
5 to 15 Minute Service Area Buffer Zones		
Stage of Breast Cancer	Number of Patients	Percent of Patients
0	91	22.52
1	188	46.53
2	81	20.05
3	26	6.44
4	10	2.48
Unknown	8	1.98
Total Patients 15 Mins	404	100.00
15 to 30 Minute Service Area Buffer Zones		
Stage of Breast Cancer	Number of Patients	Percent of Patients
0	89	21.98
1	164	40.49
2	86	21.24
3	33	8.15
4	15	3.70
Unknown	18	4.44
Total Patients 30 Mins	405	100.00
30 to 60 Minute Service Area Buffer Zones		
Stage of Breast Cancer	Number of Patients	Percent of Patients
0	42	21.11
1	74	37.19
2	41	20.60
3	21	10.55
4	8	4.02
Unknown	13	6.53
Total Patients 60 Mins	199	100.00
Outside 60 Minute Service Area Buffer Zones		
Stage of Breast Cancer	Number of Patients	Percent of Patients
0	27	18.24
1	58	39.19
2	39	26.35
3	10	6.76
4	11	7.43
Unknown	3	2.03
Total Patients 60+ Mins	148	100.00

Measuring Health Care Access to Medically Assisted Procreation in France

L'Agence de la biomédecine (Biomedicine Agency)

La Plaine Saint-Denis, France
By Florian Bayer

Contact
Florian Bayer
florian.bayer@biomedecine.fr

Software
ArcGIS 10.2 for Desktop, ArcGIS Spatial and Network Analyst, Adobe Illustrator CS5

Data Sources
L'Agence de la biomédecine 2013, Programme de Médicalisation des Systèmes d'Information 2012, Institut Géographique National 2012

In France in 2011, 45,200 women made an oocyte (a cell from which an egg or ovum develops) retrieval for a medically assisted procreation (MAP). This represents 2 percent of the national births. A limited number of MAP centers have been allocated by regions. L'Agence de la biomédecine is in charge of, among other things, measuring MAP's health-care needs in France. The objective of this study was to measure the health-care needs of MAPs and the adequacy of MAP health care in France. Another objective of the study was to offer health-care decision makers specific geographical tools. In this study, L'Agence de la biomédecine located every woman involved in a MAP from 2010 to 2011. The agency also calculated the access time to the nearest MAP center of each patient. Results show the nationwide coverage of the MAP health-care program is good, with some exception on islands and in rural territories. All maps, figures, and commentaries are included in the *Atlas of the Medically Assisted Procreation*, which assists health-care decision makers.

Courtesy of L'Agence de la biomédecine.

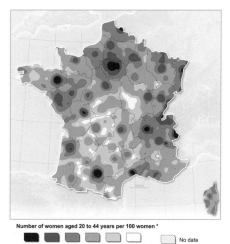

Number of women aged 20 to 44 years per 100 women *

45 36 33 30 26 23 17   No data

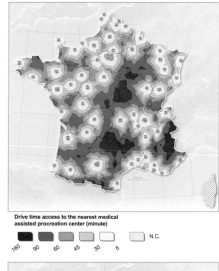

Drive time access to the nearest medical assisted procreation center (minute)

160 90 60 45 30 5   N.C.

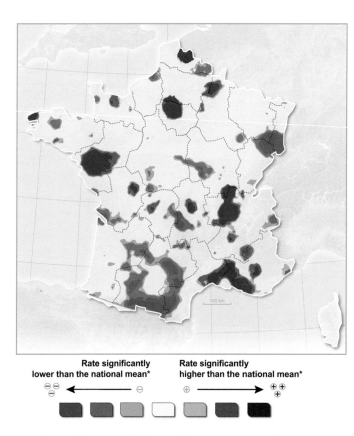

Rate significantly lower than the national mean* Rate significantly higher than the national mean*

Distribution of women who have made an oocyte retrieval in 2011

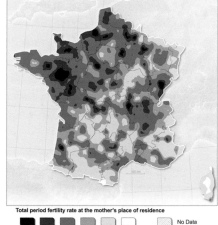

Total period fertility rate at the mother's place of residence

2.9 2.6 2.4 2.1 1.9 1.0 0.6   No Data

Regional Differences in Japan's Drinking Culture

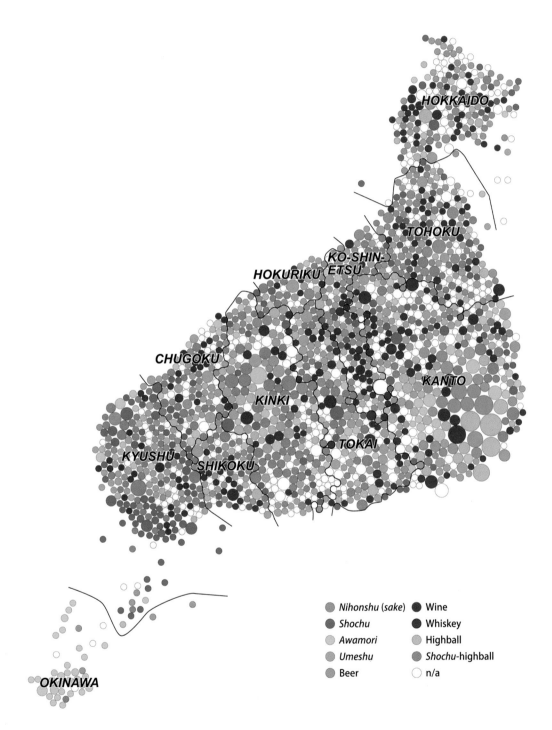

HOKKAIDO

TOHOKU

KO-SHIN-ETSU

HOKURIKU

CHUGOKU

KANTO

KINKI

TOKAI

KYUSHU

SHIKOKU

OKINAWA

- Nihonshu (sake)
- Shochu
- Awamori
- Umeshu
- Beer
- Wine
- Whiskey
- Highball
- Shochu-highball
- n/a

The University of Tokyo

Kashiwa, Chiba, Japan
By Takashi Kirimura

Contact
Takashi Kirimura
kirimura@csis.u-tokyo.ac.jp

Software
ArcGIS 10.2 for Desktop, GeoDa

Data Source
Twitter feeds

Geotagged Twitter tweets can contain substantial geographic information regarding local cultures. This map, a circle cartogram constructed using ArcGIS and GeoDa softwares, shows regional differences that were revealed by analyzing tweets that mentioned alcoholic beverages in Japan. This map was created as an example of how cultural information can be obtained and visualized by using geotagged tweet data. Each index of alcoholic beverages was calculated for each municipality based on the number of tweets mentioning each of nine types of drinks. In the circle cartogram, circle size is based on the number of all tweets in each municipal unit. The color of each circle shows the drink that had the highest index value within that municipality. If, for any of the nine types, the index is less than 100, it is marked "n/a." As the cartogram and the maps illustrate, the number of tweets for particular drinks was large in regions that are famous for the production of such drinks. Areas that exhibit high index values for *nihonshu* (sake) are in northern Japan, a major rice-producing area. Areas that exhibit high index values for *awamori* (distilled rice liquor) are in Okinawa, where *awamori* originates. In this way, local cultures—such as local drinking cultures—can be visualized by using geotagged data.

HISTORICAL AND CULTURAL

Distribution of the Sengen Shrines and Viewshed from the Top of Mount Fuji

Tokyo Map Research Company, Ltd.

Fuchu City, Tokyo, Japan
By Kei Sakamoto

Contact
Kei Sakamoto
sakamoto@t-map.co.jp

Software
ArcGIS 10.1 for Desktop

Data Sources
Foundation Geographical Data 50m Digital Elevation Model, Foundation Geographical Data 250m Digital Elevation Model, Esri Data and Maps, Map Package Lite, Point Data of Shrines in Japan

Many of the shrines bearing the name *Sengen* are related to worshiping Mount Fuji, the highest mountain/ volcano in Japan. It is thought that Sengen shrines are distributed in places where Mount Fuji can be viewed. This map is a study of the relationship between the distributed point data of Sengen shrines throughout Japan and viewshed data from the top of Mount Fuji. The Geographical Survey Institute issued a digital map of the point data representing Shinto shrines—about 50,000 points. According to this map, there are 288 Sengen shrines, and most of them are concentrated in the Kanto region and around Mount Fuji. Viewshed data from the top of Mount Fuji shows that 198 of the 288 points (68.8 percent) of Sengen shrines exist in the viewshed. Thus, the volcano can be seen from these points. Sengen shrines outside the viewshed aim at other mountains for worship or are built for the people who live in places far away from Mount Fuji.

Courtesy of Tokyo Map Research Company, Ltd.

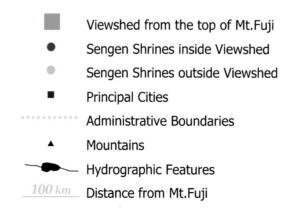

■ Viewshed from the top of Mt.Fuji

● Sengen Shrines inside Viewshed

● Sengen Shrines outside Viewshed

■ Principal Cities

.... Administrative Boundaries

▲ Mountains

━● Hydrographic Features

100 km Distance from Mt.Fuji

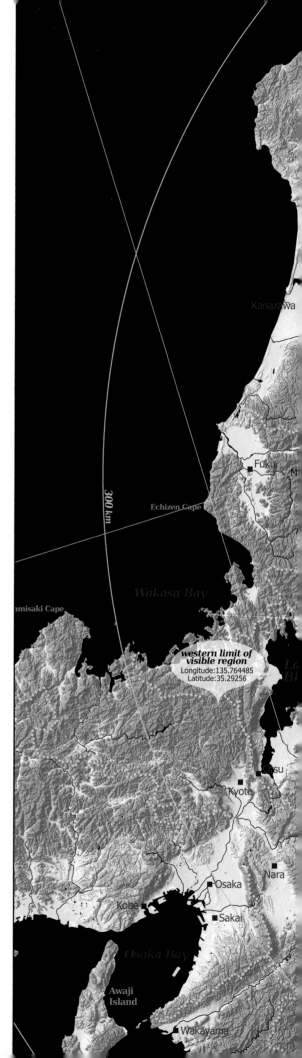

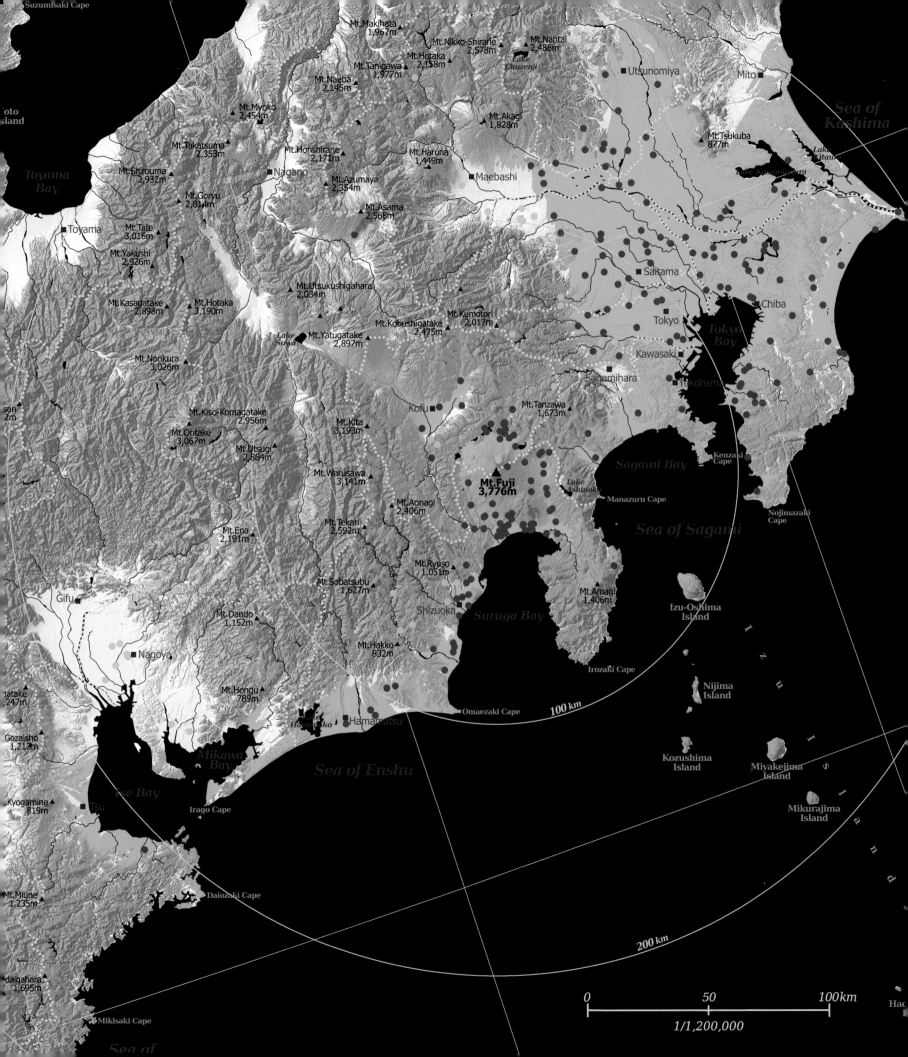

The Shape of Japan by Names

Ritsumeikan University

Kyoto, Kyoto, Japan
By Tomoki Nakaya, Keiji Yano, James Cheshire,
and Paul Longley

Contact
Tomoki Nakaya
nakaya@lt.ritsumei.ac.jp

Software
ArcGIS 10.1 for Desktop, Adobe Illustrator CC 2014 (R3.1.0)

Data Source
An original dataset compiled by Acton Wins Company, Ltd.

The geography of names provides interesting insights into Japanese origins and ancestry. For example, mapping surnames that are unique to Okinawans (people living in the southern islands of Japan) vividly depicts the cumulative outcome of their migration to other parts of the country following Japanese modernization in the mid-nineteenth century. Geographic distributions of surnames thus record the cumulative history of migration and the intermixing of people between regions. This map arises out of research to explore geographic variations in the incidence of different surnames across Japan. The dataset of surnames was created by bringing together information from detailed residential maps with nameplates and records from telephone directories as of 2007. The number of records in the underlying dataset is approximately 45 million, or roughly 90 percent of the total number of households in Japan. Word clouds were used to populate the map of Japan with the characters of surnames. The coloration of surnames corresponds to their prefectures of current residence. The size of the characters represents the number of surname occurrences in the prefecture. The borders of the prefecture are distorted using a cartometric transformation (cartogram) that renders the area of each prefecture directly proportional to its population size allowing surnames to be plotted on prefectures that are high in population but small in area.

Courtesy of Ritsumeikan University Department of Geography.

Size legend

Name	80,000	(Households)
Name	40,000	
Name	20,000	
Name	10,000	
Name	5,000	

Colour legend

Language Segregation in US Metro Areas

US Census Bureau

Washington, DC, USA
By Tiffany Julian

Contact
Meade Turner
zachariah.meade.turner.iii@census.gov

Software
ArcGIS for Desktop

Data Source
US Census Bureau

The 2011 Language Mapper is an online map pinpointing the wide array of languages spoken in homes across the nation, along with a detailed report on rates of English proficiency and the growing number of speakers of other languages. It shows where people speaking specific languages other than English live, with dots representing how many people speak each of fifteen different languages (six shown here). For each language, the mapper shows the concentration of those who report that they speak English less than "very well," a measure of English proficiency. The tool uses data collected through the American Community Survey from 2007 to 2011. This map makes it easier to serve communities. For example, businesses with this information can tailor communications to meet their customers' needs. Emergency responders can ensure they can communicate with people who need help. Schools and libraries can offer courses to improve English proficiency and offer materials written in other languages.

Courtesy of US Census Bureau.

Multiple Concentrations
Two or more areas in tight circles or bands

One side of town
Almost all dots one side of an imaginary line

Combination
One or more concentrations but many dots spread out in other areas

Single Concentration
Most dots in a single area on map

• = 100 Speakers

Los Angeles

Spanish
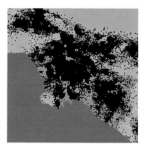

Dissimilarity Index
0.538

Population Size
4,331,266

Chinese

Pattern:
MULTIPLE

Dissimilarity Index
0.653

Population Size
399,481

Tagalog

Pattern:
COMBINATION

Dissimilarity Index
0.557

Population Size
266,890

Vietnamese

Pattern:
MULTIPLE

Dissimilarity Index
0.719

Population Size
232,912

Russian

Pattern:
ONE SIDE OF TOWN

Dissimilarity Index
0.736

Population Size
53,871

French

Pattern:
SINGLE

Dissimilarity Index
0.539

Population Size
45,130

New York   Chicago   Atlanta   Seattle

New York

Pattern:
MULTIPLE

Dissimilarity Index
0.547

Population Size
3,412,140

Pattern:
MULTIPLE

Dissimilarity Index
0.683

Population Size
549,120

Pattern:
MULTIPLE

Dissimilarity Index
0.662

Population Size
137,942

Pattern:
MULTIPLE

Dissimilarity Index
0.882

Population Size
24,623

Pattern:
MULTIPLE

Dissimilarity Index
0.732

Population Size
240,841

Pattern:
ONE SIDE OF TOWN

Dissimilarity Index
0.538

Population Size
135,816

Chicago

Pattern:
MULTIPLE

Dissimilarity Index
0.573

Population Size
1,482,217

Pattern:
COMBINATION

Dissimilarity Index
0.707

Population Size
77,460

Pattern:
COMBINATION

Dissimilarity Index
0.637

Population Size
74,196

Pattern:
MULTIPLE

Dissimilarity Index
0.864

Population Size
18,056

Pattern:
ONE SIDE OF TOWN

Dissimilarity Index
0.767

Population Size
39,186

Pattern:
ONE SIDE OF TOWN

Dissimilarity Index
0.594

Population Size
28,286

Atlanta

Pattern:
COMBINATION

Dissimilarity Index
0.462

Population Size
33,394

Pattern:
ONE SIDE OF TOWN

Dissimilarity Index
0.675

Population Size
33,394

Dissimilarity Index
0.764

Population Size
6,247

Pattern:
MULTIPLE

Dissimilarity Index
0.748

Population Size
33,763

Pattern:
ONE SIDE OF TOWN

Dissimilarity Index
0.778

Population Size
11,676

Dissimilarity Index
0.457

Population Size
30,457

Seattle

Dissimilarity Index
0.390

Population Size
194,495

Pattern:
COMBINATION

Dissimilarity Index
0.582

Population Size
67,963

Dissimilarity Index
0.524

Population Size
38,771

Pattern:
COMBINATION

Dissimilarity Index
0.611

Population Size
44,700

Dissimilarity Index
0.581

Population Size
28,336

Pattern:
SINGLE

Dissimilarity Index
0.484

Population Size
14,825

Where Does the Food on MyPlate Come From?

US Department of Agriculture (USDA) National Agricultural Statistics Service

Washington, DC, USA

By Alissa Bartholomew, Melanie Edwards, Aleksey Minchenkov, Amanda Morris, Ray Roberts, and Krissy Young

Contact
Ray Roberts
Ray.Roberts@nass.usda.gov

Software
ArcGIS 10.1 for Desktop

Data Source
USDA National Agricultural Statistics Service

To help young minds better understand where the food on their plate comes from, USDA's National Agricultural Statistics Service (NASS) linked 2012 Census of Agriculture data and USDA's MyPlate initiative. Using maps to display the most recent Census of Agriculture results, NASS shows where foods in the five main food groups—dairy, fruits, grains, proteins, and vegetables—are grown in the United States. By allowing students to explore US agriculture production and the food they eat, this tool provides them with development opportunities for agriculture, nutrition, math, and science.

NASS combined county-level agricultural production data to provide the most local level data possible. Students can use the maps in combination with the most recent census data to explore what most commonly eaten foods are grown in their own state and county. They can also analyze data, learn information such as the local acreage dedicated to fruits, and analyze the availability of all food groups in their area.

Courtesy of USDA National Agricultural Statistics Service.

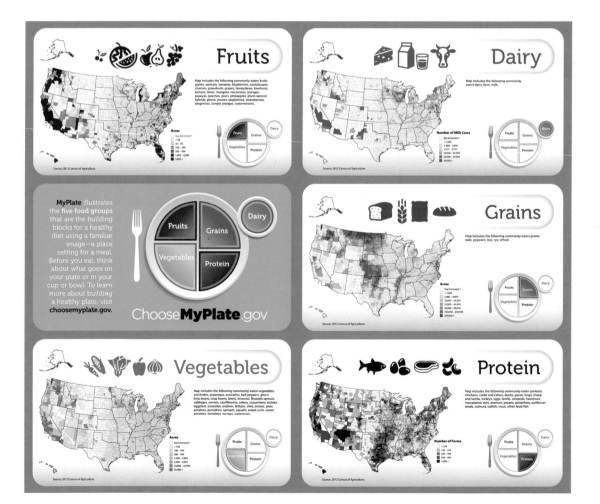

Demographic Trends among Wyoming Farm and Ranch Operators, 1920–2007

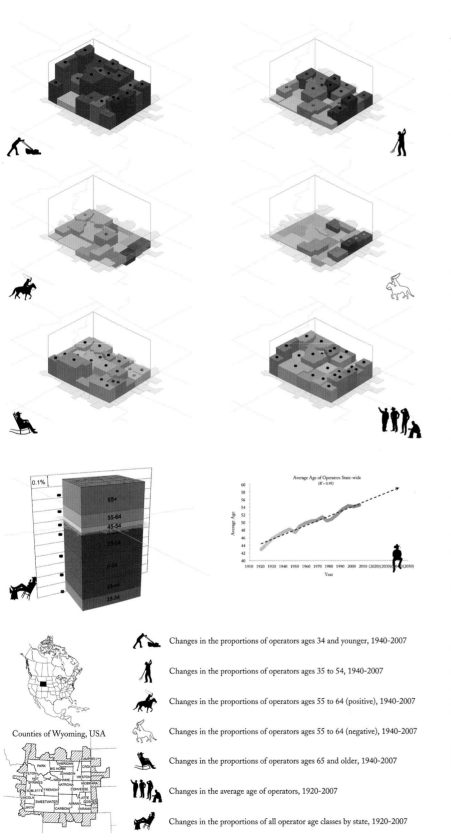

Yale University School of Forestry and Environmental Studies

New Haven, Connecticut, USA

By Henry B. Glick, Lindsi Seegmiller, Charlie Bettigole, Ambika Khadka, and Chadwick D. Oliver

Contact
Henry B. Glick
henry.glick@gmail.com

Software
ArcGIS 10.1 for Desktop, ArcGIS 3D Analyst, ArcScene 10.1, RStudio v. 0.97.551, Microsoft Excel 2010, Inkscape v. 0.48.4

Data Source
US Census of Agriculture 1920 through 2007

Like other portions of the United States, Wyoming's agricultural community has experienced dramatic shifts since the early twentieth century. One of these shifts is an aging demographic. Over time, fewer and fewer young operators have taken up land-based occupations. These maps illustrate changes in the average age and the relative proportions of farm and ranch operator age classes. They provide a means of distilling large amounts of statistical and tabular data into select visuals, offering critical information in a digestible format. This study provides useful context for strengthening ecological research, outreach, and communication among farming and ranching communities in the High Plains of the western United States.

Courtesy of Yale University School of Forestry and Environmental Studies.

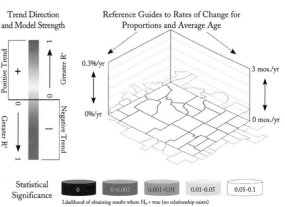

Counties of Wyoming, USA

Changes in the proportions of operators ages 34 and younger, 1940-2007

Changes in the proportions of operators ages 35 to 54, 1940-2007

Changes in the proportions of operators ages 55 to 64 (positive), 1940-2007

Changes in the proportions of operators ages 55 to 64 (negative), 1940-2007

Changes in the proportions of operators ages 65 and older, 1940-2007

Changes in the average age of operators, 1920-2007

Changes in the proportions of all operator age classes by state, 1920-2007

The Changing Face of Agriculture on Maui

County of Maui

Wailuku, Hawaii, USA
By Jon Gushiken and Nancy Swienton

Contact
Jon Gushiken
jon.gushiken@co.maui.hi.us

Software
ArcGIS 10.2.1 for Desktop

Data Source
Maui County real property tax data

The changing quantity of acreage dedicated to the four main classes of agricultural land use on Maui Island—diversified, pineapple, sugar cane, and pasture—reflects the changing needs of consumers and industry over the past fifteen years. Since 1999, the most significant decreases were in the mostly commercial uses of pasture and pineapple. Diversified use, more concentrated on residential parcels among private landowners, has remained relatively static. Sugar cane classification has also remained stable but may decrease as more development occurs on the island's central isthmus.

Courtesy of County of Maui Finance Department, Real Property Assessment Division; County of Maui Planning Department.

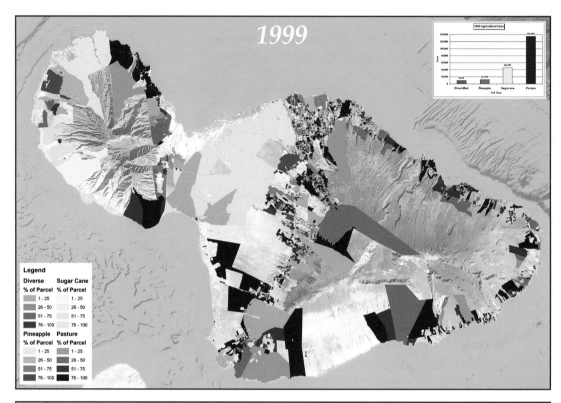

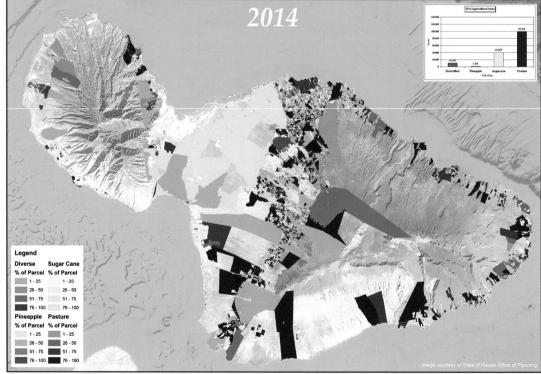

Biodiversity and Natural Communities of the National Forests in Florida

US Department of Agriculture (USDA) Forest Service

Tallahassee, Florida, USA
By Paul B. Medley
Contact
Paul B. Medley
pbmedley@fs.fed.us

Software
ArcGIS 10.1 for Desktop

Data Sources
Critical Lands and Water Identification Project (CLIP) v.3.0.; Cooperative Land Cover (CLC) v.2.3

National forests span more than 1.2 million acres in north and central Florida and are essential havens for rare plant and animal species. For example, the Ocala National Forest is home to the largest population of threatened Florida scrub-jays and the Apalachicola National Forest has the largest population of endangered red-cockaded woodpeckers. In fact, the panhandle region of Florida where the Apalachicola is located has been identified as one of North America's five hotspots for species rarity and richness. At least 127 rare species of plants and vertebrates and forty-five of the sixty-two terrestrial communities in Florida are found in this region. Florida's national forests contain a broad assortment of natural community types, including pine flatwoods, wet prairies, and a large swath of oak scrub in the Ocala National Forest. To highlight focal communities of Florida's national forests, the inset maps show the array of natural community types on each forest. The biodiversity resource priorities map illustrates the biodiversity priority areas in Florida, highlighting the importance of the national forests for protecting these vital resources.

Courtesy of USDA Forest Service, National Forests in Florida.

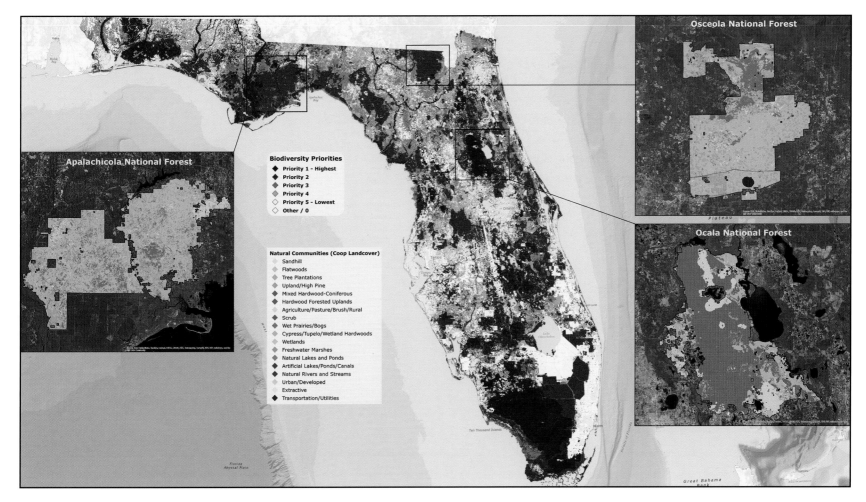

Fort Belknap Indian Reservation Forestry and Fire Planning Map

Bureau of Indian Affairs, Rocky Mountain Region

Billings, Montana, USA
By Kevin Nelstead

Contact
Kevin Nelstead
kevin.nelstead@bia.gov

Software
ArcGIS 10.1 for Desktop

Data Sources
Bureau of Indian Affairs, Montana State Library, US Department of Agriculture Natural Resources Conservation Service, US Bureau of Land Management

Fort Belknap Indian Reservation in north central Montana needed a map for the forestry and fire planning and fieldwork activities of the Bureau of Indian Affairs (BIA). The map shown here filled the need with features that are important for forest development, hazardous fuels reduction projects, and wildland fire suppression. Natural features, such as topography and hydrography, are the map's foundation. BIA transportation data was out of date, having been extracted from forty-year-old US Geological Survey topographic quadrangles. Starting with the Montana Transportation Framework data, the entire road network was reviewed and edited based on US Department of Agriculture aerial imagery. Some of the landmark features on this map, such as flight obstructions, were added because of their importance for aviation operations, while others, such as grain bins, were added as landmarks on the prairies where few other visible landmarks exist. Forestry and fire personnel are using a digital version of this map on mobile devices in the field. This map is the companion product to a ninety-page map book of the 622,000-acre Fort Belknap Indian Reservation. Similar products for all seven reservations in the Rocky Mountain Region of the BIA are planned.

Courtesy of Bureau of Indian Affairs, Rocky Mountain Region.

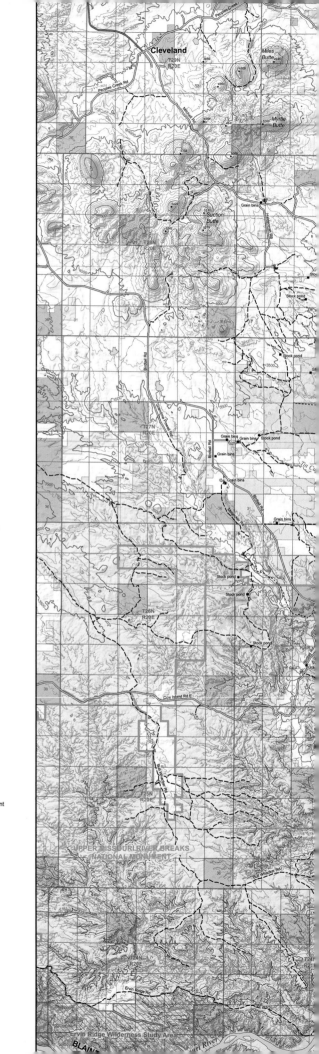

0	5	10	15

Miles

1:100,000

Mapping Forest Ownership Types across the Conterminous United States

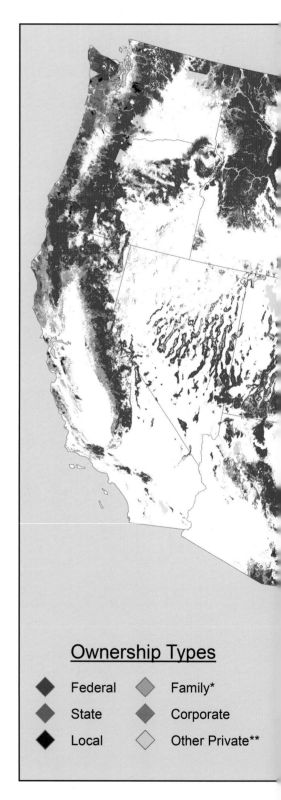

Family Forest Research Center, US Department of Agriculture (USDA) Forest Service, and University of Massachusetts

Amherst, Massachusetts, USA

By Jaketon H. Hewes, Brett J. Butler, Greg C. Liknes, Mark D. Nelson, and Stephanie A. Snyder

Contact
Jaketon H. Hewes
jhewes@eco.umass.edu

Software
ArcGIS 10.0 for Desktop

Data Sources
Forest Inventory and Analysis program plot data, Protected Areas Database of the United States

Who makes decisions about using forest resources in the United States? In some cases, land is publicly owned, and may be used in a way that balances the needs and desires of the public with broader conservation objectives. In other cases, private companies own the land and produce jobs and profits. Yet, contrary to public perception, individuals and families, in aggregate, are the biggest decision-making group of forestland owners. This map illustrates the spatial arrangement of forest ownership in the United States. Ownership is portrayed in six categories. Public ownership categories include federal, state, and local. Private ownership categories include family, corporate, and other private. Family forest ownership comprises 35 percent of the total forestland and is the predominant owner type in the eastern United States. Federal ownership dominates in the western United States and comprises 32 percent of all forestland. The map highlights a conservation challenge because much of the forestland is controlled by thousands of owners, each with different values and intentions for their land. Understanding the spatial distribution of forest ownership types across the country helps policy makers, researchers, conservation organizations, and other interested groups identify where to focus programs and policies related to conservation and timber production, and enhance outreach and education efforts.

Courtesy of Jaketon Hewes, Family Forest Research Center.

Ownership Types

◆ Federal   ◆ Family*

◆ State   ◆ Corporate

◆ Local   ◇ Other Private**

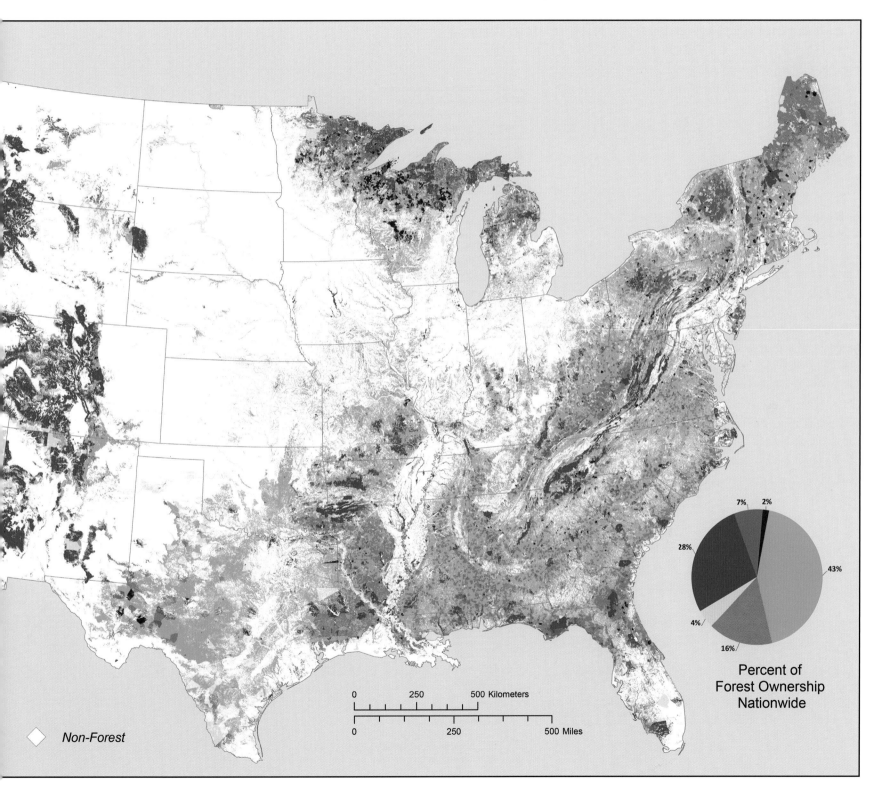

Non-Forest

0   250   500 Kilometers

0   250   500 Miles

7%   2%

28%

43%

4%

16%

Percent of
Forest Ownership
Nationwide

Trinity Alps Wilderness

US Department of Agriculture (USDA) Forest Service

Vallejo, California, USA

By Jean Ann Carroll, Daniel Spring,
and Don Van Nice

Contact
Rich Spradling
rspradling@fs.fed.us

Software
ArcGIS 10.0 for Desktop, Adobe Illustrator

Data Source
USDA Forest Service

This map shows the Trinity Alps Wilderness at a scale of 1 inch to the mile with 80-foot contours. The map was published in 2013 by the Forest Service's Pacific Southwest Regional Office. The wilderness area covers 525,627 acres (2,130 square kilometers) in Northern California, roughly between Eureka and Redding. Designed for foot and horse travel within the wilderness, the map includes trails, recreation sites, land management and surrounding roads. Shaded relief and vegetation cover are also illustrated.

Courtesy of USDA Forest Service.

	National Forest Boundary
	Wilderness Boundary
	National Recreation Area Boundary
	Special Area Boundary
	Ranger District Boundary
	Forest Service Land
BLM	Bureau of Land Management Land
	Non Federal Lands
	Wild and Scenic River
	Highway
	Paved Road
	Gravel Road	(Suitable for Passenger Cars)
	Dirt Road
	Road not maintained for passenger cars
4WD	Road not maintained for passenger cars -4 wheel drive recommended
OHV	Designated Off-Highway Vehicle Route
	National Scenic Byway
	National Scenic Trail
	National Recreation Trail
	Maintained Trail
	Maintained Trail-not recommended for horses
	Unmaintained Trail
(3)	State Highway
(211)	County Road
(17)	Primary Forest Route
37N56	Forest Road Number
8W19	Forest Trail Number
	Forest Service Ranger District Office
	Other Forest Service Facility
	Forest Service Campground
	Group Campground
	Forest Service Day Use Area
	Campground, Non-Forest Service
	Picnic Area, Non-Forest Service
	Visitor Information Center
	Trailhead
	Equestrian Trail
	Historical Site
	View Point
	Wildlife Viewing Area
	Trailer Dump Station
	Boat Ramp
	Beach
	Marina
	Resort
	Corral
	Mine
	Spring
	Locked Gate
	Located Landmark
	Horizontal Control Station
	Building Site
	Helitack Base

	Perrenial Stream
	Intermittent Stream
	Lake, Reservoir, or Large River
	Marsh or Swamp
	Permanent Snowfield
18	Section Line and Section Number
R. 7 E. / T. 6 N.	Township Line and Township & Range Number
CALLAHANS	1:24,000 scale USGS/USFS Quadrangle Map Name and Area of Coverage

Vegetation

Woodland   Scrub

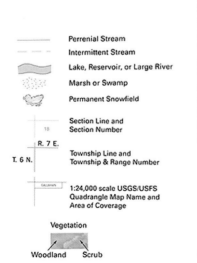

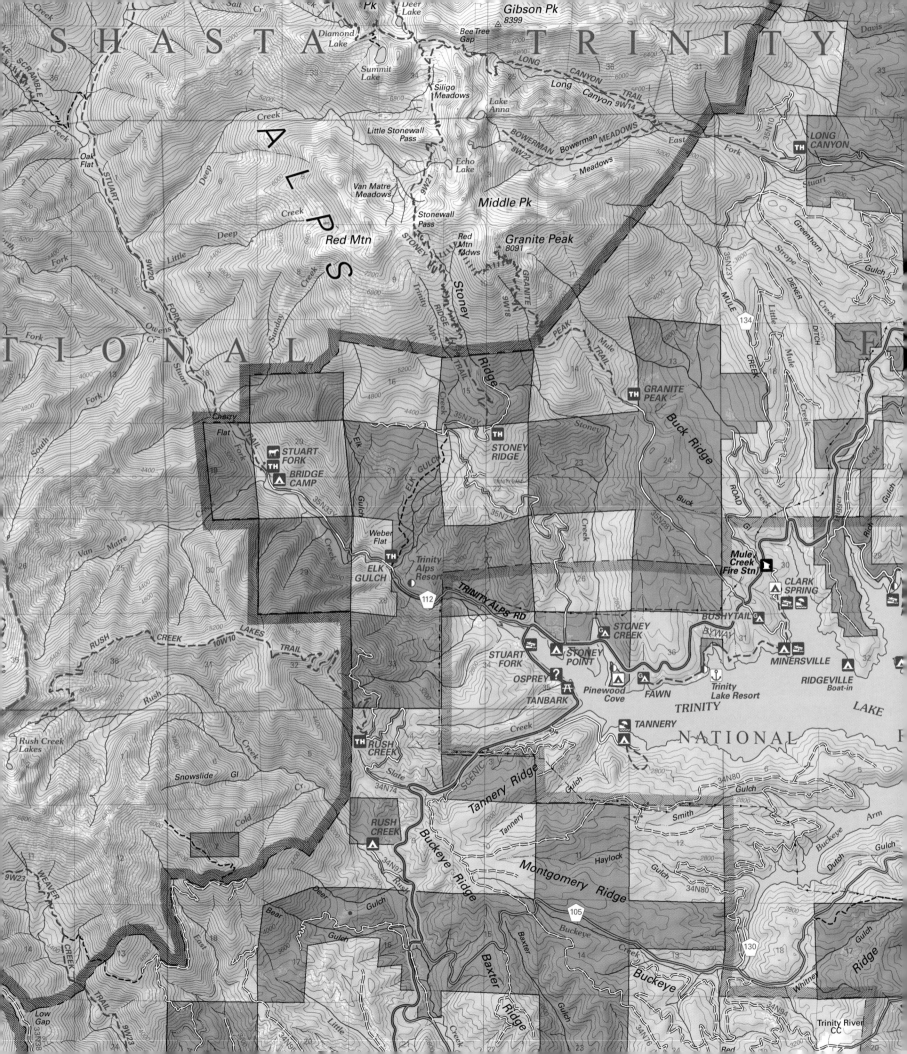

Metallic and Industrial Minerals in Alberta— Concentrations of MIM Permits

Alberta Department of Energy

Edmonton, Alberta, Canada
By Natasha Clarke, Hanna Oh, and
John Davies

Contact
Stephane La Rochelle
stephane.larochelle@gov.ab.ca

Software
ArcGIS 10.1 for Desktop, Microsoft Excel
2010, Adobe Photoshop 9.0, Adobe
Illustrator 12.0.0

Data Source
Alberta Department of Energy

The Alberta Department of Energy's Metallic and Industrial Mineral (MIM) permit data was processed to generate a series of density grid maps. The highest density areas (hot spots) indicate high concentrations of MIM (exploration) permit holdings from 1955, when MIM permits were established in Alberta, to the present day. In addition, data was extrapolated on these densities to create a predictive map of likely areas for future MIM exploration. The Coal and Mineral Development business unit within the department uses the predictive map to guide decisions.

Courtesy of Alberta Department of Energy.

Modernizing the US Vertical Datum: Components of the Geoid

National Oceanic and Atmospheric Administration

Silver Spring, Maryland, USA
By Monica Youngman

Contact
Monica Youngman
Monica.Youngman@noaa.gov

Software
ArcGIS 10.2 for Desktop

Data Source
National Oceanic and Atmospheric Administration's
National Geodetic Survey

The mission of the National Oceanic and Atmospheric Administration's National Geodetic Survey (NGS) is to define, maintain, and provide access to the National Spatial Reference System. NGS plans to replace the official datums in 2022 using modern technology and methods to address known systematic issues. The vertical datum will use newly collected gravity data obtained through the Gravity for the Redefinition of the American Vertical Datum (GRAV-D) project. Because gravity determines where water will flow, it is critical to incorporate accurate gravity data into a height measurement system. Although measurements can be taken from any surface (such as an ellipsoid), to have accurate orthometric heights (roughly heights above mean sea level) it is critical to have a highly accurate picture of the gravity field.

The GRAV-D project is collecting airborne gravity data covering the United States and territories to provide information about mid-sized features not captured by satellite and terrestrial data. Similar to aerial photogrammetry, the measurements at various distances from Earth show different levels of detail. The airborne gravity campaign, along with satellite and surface gravity datasets, are used to calculate the geoid, which most closely aligns with mean sea level. The geoid is used as the "zero" surface for the vertical datum from which orthometric heights are measured.

Courtesy of Monica Youngman, NOAA.

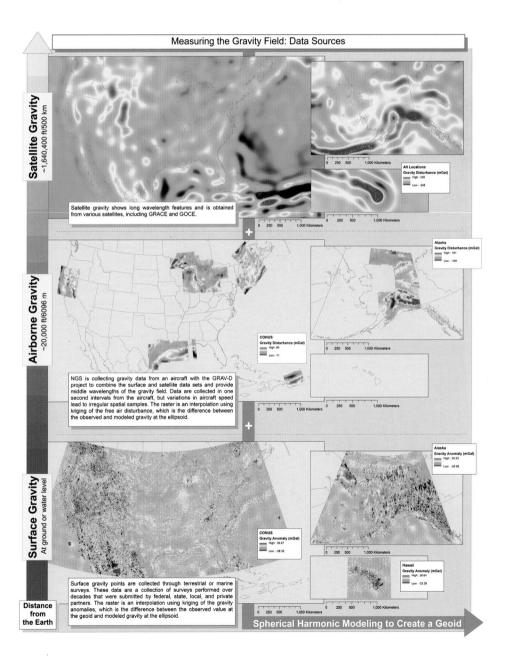

Geologic Map of the Granite 7.5-Minute Quadrangle, Lake and Chaffee Counties, Colorado

US Geological Survey

Denver, Colorado, USA

By Ralph R. Shroba, Karl S. Kellogg, and Theodore R. Brandt

Contact

Theodore Brandt

tbrandt@usgs.gov

Software

ArcGIS for Desktop, Global Mapper, MAPublisher

Data Source

US Geological Survey

This map portrays the geology within an area of about 150 square kilometers (58 square miles) near the town of Granite, Colorado. The diverse landscape of the map area is due in part to surficial and tectonic processes during the Quaternary Period and Neogene Subperiod of geologic time. The area is popular with outdoor enthusiasts attracted by spectacular mountain scenery, streams, reservoirs, and hiking trails in and near the San Isabel National Forest. The distribution of surficial deposits and the dry union formation shown on the geologic map (with and without shaded relief) was determined with the aid of lidar data that covers about 88 percent of the quadrangle, a 10-foot contour map generated from the lidar data, and imagery from the National Agriculture Imagery Program.

Courtesy of US Geological Survey.

SURFICIAL DEPOSITS

Manmade deposits

af	Artificial-fill deposits (latest Holocene)
pt	Placer-tailings deposits (latest Holocene)
mt	Mine-tailings deposits (latest Holocene)

Lacustrine deposits

| Qb | Beach deposits (latest Holocene) |

Alluvial deposits

Qva	Valley-floor alluvium (Holocene and latest Pleistocene?)
Qsw	Sheetwash alluvium (Holocene and late Pleistocene)
Qsw/Nd	Sheetwash alluvium over Dry Union Formation

Outwash of Pinedale age (late Pleistocene)

| Qopy | Younger unit |
| Qopo | Older unit |

Outwash of Bull Lake age (late and middle Pleistocene)

| Qoby | Younger unit (late and middle Pleistocene) |
| Qobo | Older unit (middle Pleistocene) |

Outwash of pre-Bull Lake age (middle Pleistocene)

| Qoey | Younger unit |
| Qobeo | Older unit |

Alluvial and mass-movement deposits

| Qac | Alluvium and colluvium, undivided (Holocene to middle? Pleistocene) |
| Qf | Fan deposits (Holocene to middle? Pleistocene) |

Mass-movement deposits

Qta	Talus deposits (Holocene to middle? Pleistocene)
Qc	Colluvium (Holocene to middle? Pleistocene)
Qc/Nd	Colluvium over Dry Union Formation
Qls	Landslide deposits (Holocene to middle? Pleistocene)

Glacial deposits

| Qrg | Rock-glacier deposits (early Holocene? and latest Pleistocene?) |

Till of Pinedale age (late Pleistocene)

Qtpy	Younger unit
Qtpo	Older unit
Qtb	Till of Bull Lake age (late and middle Pleistocene)
Qte	Till of pre-Bull Lake age (middle Pleistocene)

BEDROCK UNITS

Post-volcanic sediments

| Nd | Dry Union Formation (lower Pliocene? and Miocene) |

Volcanic rocks

| Ɍeva | Vesicular andesite flow (upper Oligocene) |

Hypabyssal rocks

Ɍerc	Hypabyssal rhyolite of Clear Creek (late Oligocene)
Ɍedt	Microtonalite (late Oligocene?)
Ɍewf	Felsic plutons and dikes of Winfield Peak and Middle Mountain of Fridrich and others (1998) (middle Eocene)

| Ɍ⋅Kr | Rhyolite dikes (Paleocene or Late Cretaceous) |
| Ɍ⋅Ki | Felsic porphyry intrusive (Paleocene or Late Cretaceous) |

Proterozoic igneous and metamorphic rocks

YXp	Pegmatite (Mesoproterozoic and (or) Paleoproterozoic)
Ygr	Monzogranite and quartz monzonite (Mesoproterozoic)
Ygl	Monzogranite of Langhoff Gulch (Mesoproterozoic)
Yglf	Strongly foliated phase
Ygrg	Granite of Granite (Mesoproterozoic)
Xgg	Granitic gneiss (Paleoproterozoic)
Xlg	Leucogranite gneiss (Paleoproterozoic)
Xb	Biotite gneiss (Paleoproterozoic)
Xhg	Hornblende gneiss and amphibolite (Paleoproterozoic)
Xgb	Gabbro (Paleoproterozoic)

EXPLANATION

Contact—Dashed where approximately located

Fault, type unspecified—Dashed where approximately located; dotted where concealed beneath surficial deposits

Normal fault—Dashed where approximately located; short-dashed where inferred; dotted where concealed beneath surficial deposits; bar-and-ball symbol on apparent downthrown side; traces of inferred faults were identified from lidar imagery

Shoreline—Observed on lidar imagery

Fluvial scarp—Observed on lidar imagery; hachures are on riser between lower and higher surfaces and point downslope

Landslide scarp—Observed on lidar imagery and NAIP imagery; lower limit of hachures are on or near base of scarp. Prominent upslope facing scarps (sackungen) locally formed on the large landslide deposits derived from Proterozoic bedrock on the north side of Clear Creek near the southwest corner of the map area

Toe of inactive solifluction deposits—Observed on NAIP imagery; sawteeth pattern is on the upslope side. Solifluction deposits were formed by downslope flowage of surficial material in the large alpine area near the head of Lost Canyon north of Clear Creek. Approximately located

Shear zone—Marked by ductile deformation fabric in Proterozoic rocks

Crest line of moraine—Observed on lidar imagery and NAIP imagery

Lineament—Observed on lidar imagery and NAIP imagery. Linear features locally may coincide with zones of fracturing in Proterozoic bedrock, tabular intrusive bodies, or linear mass-movement deposits

Inclined bedding—Showing strike and dip

Fault attitude—Showing fault dip

Inclined foliation—Showing strike and dip. Measurements of inclined foliation in areas of Quaternary surficial deposits (units Qc, Qac, and Qopy?) were made on Proterozoic bedrock outcrops that are too small to show at map scale

Vertical foliation—Showing strike

Mineral lineation—Showing bearing and plunge, depicted by an arrowhead

Sample—Location and isotopic ages of bedrock (see table 3 in pamphlet)
GR11–725
42.25 ± 0.13 Ma

US Natural and Industrial Resources

Vermont Agency of Transportation

Montpelier, Vermont, USA
By Stephen Smith

Contact
Stephen Smith
stephen.smith@state.vt.us

Software
ArcGIS 10.2.1 for Desktop, Inkscape

Data Sources
US Geological Survey, US Energy Information Administration, Open Energy Information, Wikipedia, US Department of Agriculture, Natural Earth, Mapbox Maki

This map pays homage to a travel poster produced by the British Information Services in the early 1940s. It depicts the natural and industrial resources of the continental United States overlaid on a simplified land-use model. This map was an exercise in replicating the handmade appearance of an old grade-school wall atlas using modern data and tools. The biggest challenge was to give the illusion of smooth, well-defined land use areas such as grassland, forest, and agriculture. Closely emulating historical designs and replicating the constraints of the manual design era often result in something attractive and unique.

Courtesy of Vermont Agency of Transportation.

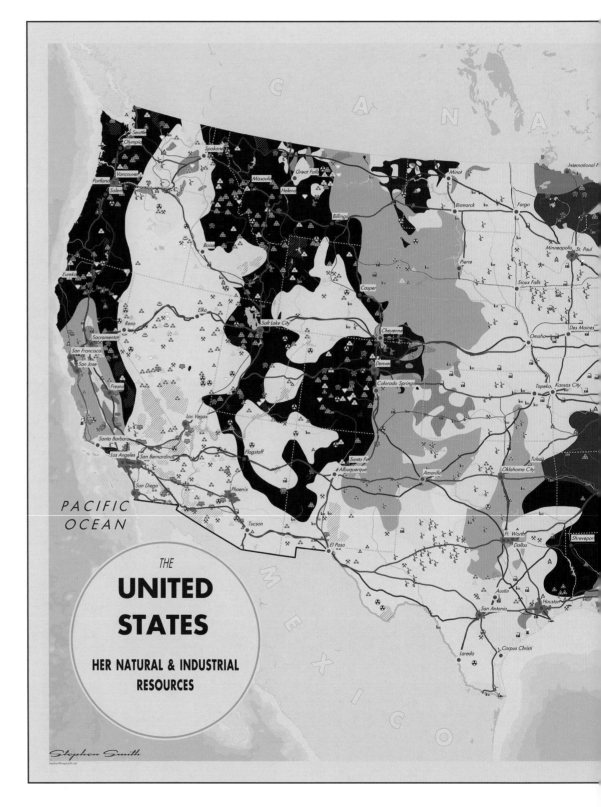

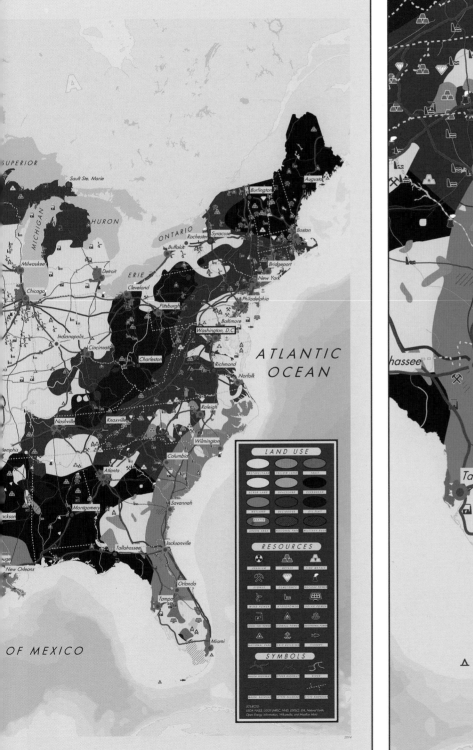
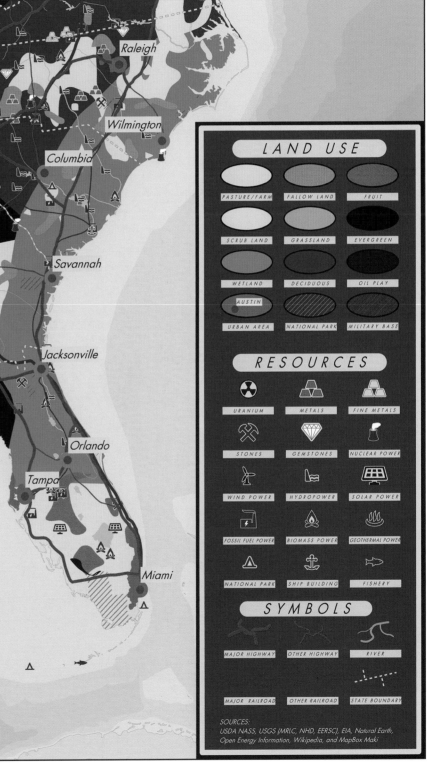

App for Esri Climate Resilience App Challenge

CLIMsystems

Hamilton, Waikato, New Zealand
By Peter Kouwenhoven

Contact
Peter Kouwenhoven
pkouwenh@climsystems.com

Software
ArcGIS for Desktop, SimCLIM for ArcGIS/
Marine

Data Source
CLIMsystems

Sea levels around the globe rise as air temperatures rise because of greenhouse gas emissions. This increases the melting of land ice and, more importantly, raises the oceans' temperatures which cause sea water to expand. Other processes play a role in sea-level rise, including drops in air pressure, changes in ocean currents, and the distribution of ice masses. Another critical and often overlooked factor in sea level rise is that land can move up or down. The magnitude of this process can be comparable to that of sea level rise. When land rises, it lowers the rate of sea level rise experienced locally. When

land sinks, it exacerbates the local sea-level rise. CLIMsystems created an app for the Esri Climate Resilience App Challenge that shows a global map of the combined processes of local (absolute) sea level rise and local vertical land movement. The sea-level rise values are based on the largest greenhouse gas emissions projected in climate modeling scenarios. The vertical land movement values were generated from direct observations through continuous GPS and from trend analysis of tidal observations.

Courtesy of CLIMsystems.

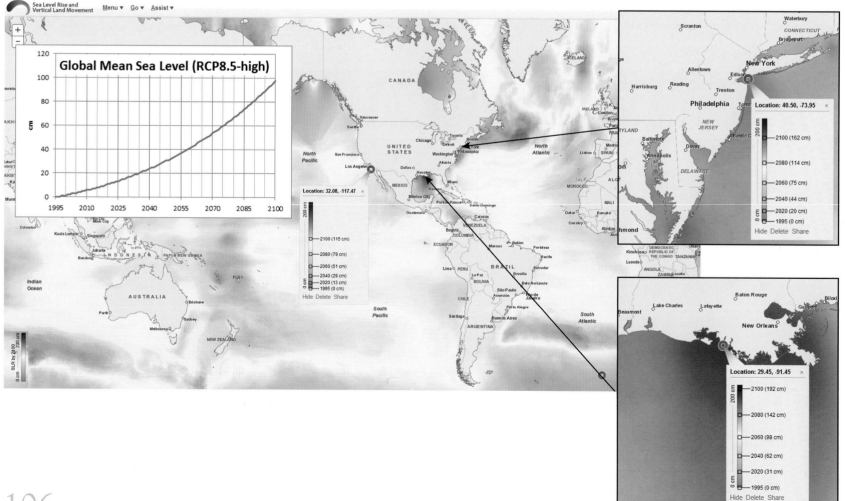

Geology of Stackhouse-Numabin Bays, Reindeer Lake Saskatchewan

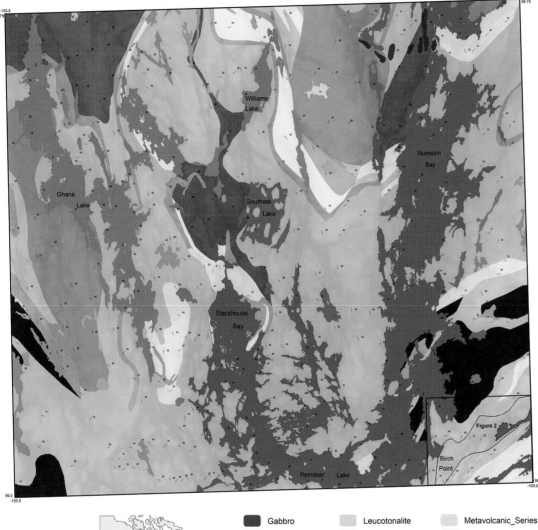

Community College of Baltimore County

Catonsville, Maryland, USA
By Audrey Garven

Contact
Scott Jeffrey
sjeffrey@ccbcmd.edu

Software
ArcGIS for Desktop

Data Source
Map based on work by author, 1975–1976

This map presents a geologically complex region located within the Precambrian Shield of northern Saskatchewan. The region is composed of ancient sedimentary, volcanic, and granitic rocks thought to be originally part of an island arc terrain. Two analytical methods were used to help visualize elements of this study area. The first involved finding and converting a digital elevation model to a raster, and then exporting the result to ArcScene where the raster was extruded and the geology map draped over the 3-D image. The second method was an attempt to visualize folded structures in the study area. The author generated numerous cross sections, creating mass points that were then used to interpolate to a raster.

Courtesy of Community College of Baltimore County.

Geologic Map of Mars

US Geological Survey

Flagstaff, Arizona, USA

By Kenneth L. Tanaka, James A. Skinner Jr., James M. Dohm, Rossman P. Irwin III, Eric J. Kolb, Corey M. Fortezzo, Thomas Platz, Gregory G. Michael, and Trent M. Hare

Contact
Trent M. Hare
thare@usgs.gov

Software
ArcGIS 10.2 for Desktop

Data Sources
National Aeroautics and Space Administration, Arizona State University, US Geological Survey

This global geologic map of Mars, which records the distribution of geologic units and landforms on the planet's surface through time, is based on unprecedented variety, quality, and quantity of remotely sensed data acquired since the Viking Orbiters mission of the 1970s. This data has provided morphologic (shape and form), topographic, spectral, thermophysical, radar-sounding, and other observations for integration, analysis, and interpretation in support of geologic mapping. In particular, the precise topographic mapping now available enables consistent morphologic portrayal of the surface for global mapping. Previously used visual-range image bases were less effective, because they combined morphologic and albedo (surface reflectivity) information and, locally, atmospheric haze. Also, thermal infrared image bases used for this map tended to be less affected by atmospheric haze and thus are reliable for analysis of surface morphology and texture at even higher resolution than the topographic products.

Tanaka, K. L., J. A. Skinner Jr., J. M. Dohm, R. P. Irwin III, E. J. Kolb., C. M. Fortezzo, T. Platz, G. G. Michael, and T. M. Hare. 2014. Geologic Map of Mars: US Geological Survey Scientific Investigations Map 3292, scale 1:20,000,000. http://dx.doi.org/10.3133/sim3292.

Courtesy of US Geological Survey and National Aeronautics and Space Administration.

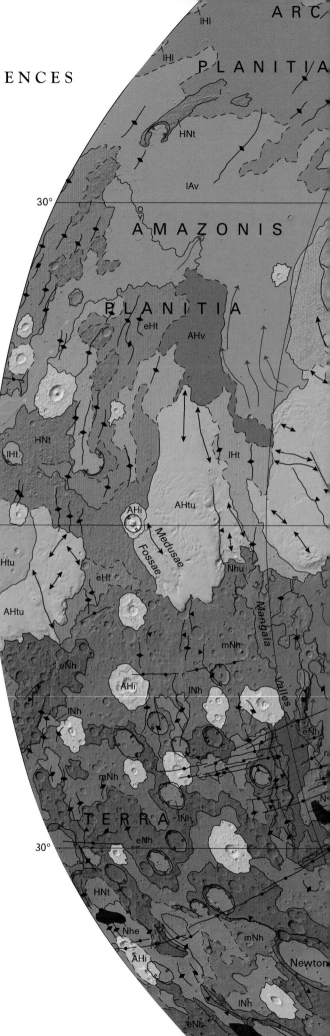

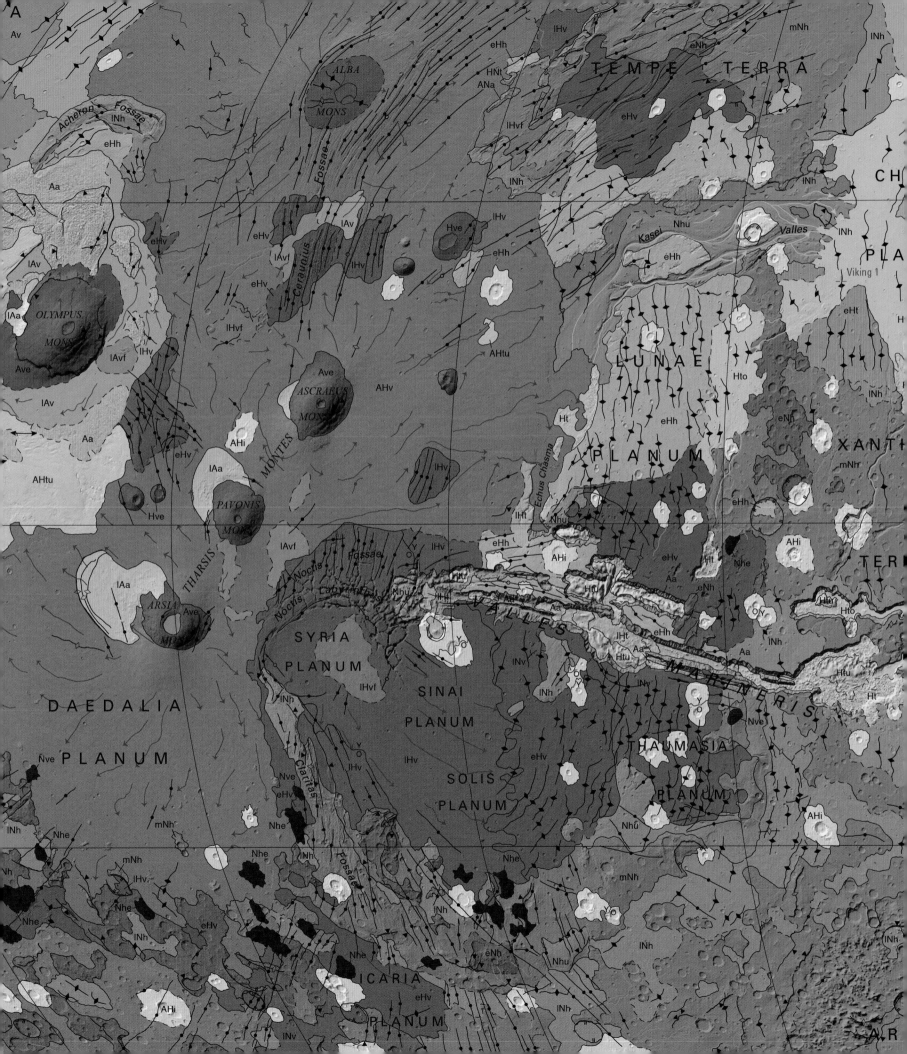

Hydraulic Fracturing and Water Stress: Water Demand by the Numbers

Blue Raster and Ceres

Arlington, Virginia, USA
By Adrienne Allegretti (Blue Raster) and
Monika Freyman (Ceres)

Contact
Adrienne Allegretti
aallegretti@blueraster.com

Software
ArcGIS 10.2 for Desktop, Adobe Illustrator

Data Sources
US Geological Survey Groundwater
Depletion data, US Drought Monitor data,
PacWest, FracDB/FracFocus.org

The drought monitor map shows hydraulically fractured wells overlaid with US drought data from January 7, 2014. Resulting statistics show that over 55 percent of wells are in regions experiencing drought conditions at that time. The groundwater depletion map shows hydraulically fractured wells overlaid with US groundwater depletion data from 1900 to 2008 in forty assessed aquifer systems. Resulting statistics for this map show that over 36 percent of wells are in regions of groundwater depletion.

Courtesy of Ceres and Blue Raster.

HYDRAULICALLY FRACTURED
oil and gas wells against groundwater depletion data

A database of hydraulically fractured wells is overlaid on a U.S. Geological Survey (USGS) map showing cumulative groundwater depletion, from 1900 through 2008, in 40 assessed aquifer systems. Over 36 percent of the 39,294 wells happen to be in regions of groundwater depletion. In the central U.S. hatched colors indicate that there are overlapping aquifers with different values of depletion. The USGS map and accompanying report can be found at http://pubs.usgs.gov/sir/2013/5079/SIR2013-5079.pdf

Well data sourced from PacWest FracDB / FracFocus.org. Well data reflects reporting of wells hydraulically fractured between 01/2011 to 05/2013.

Please see www.ceres.org for the accompanying report.

Groundwater Depletion, in cubic kilometers

-40 to -10	10 to 25
-10 to 0	25 to 50
0 to 3	50 to 150
3 to 10	150 to 400

39,294 WELLS REPORTED

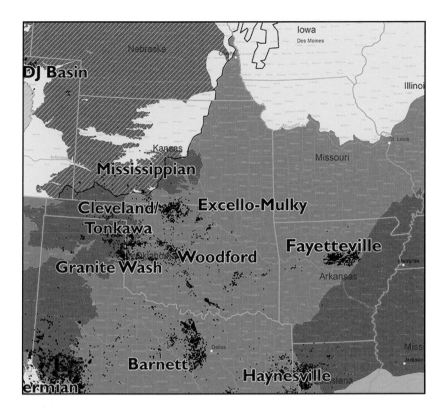

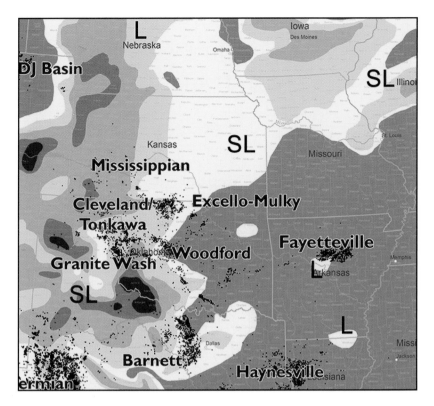

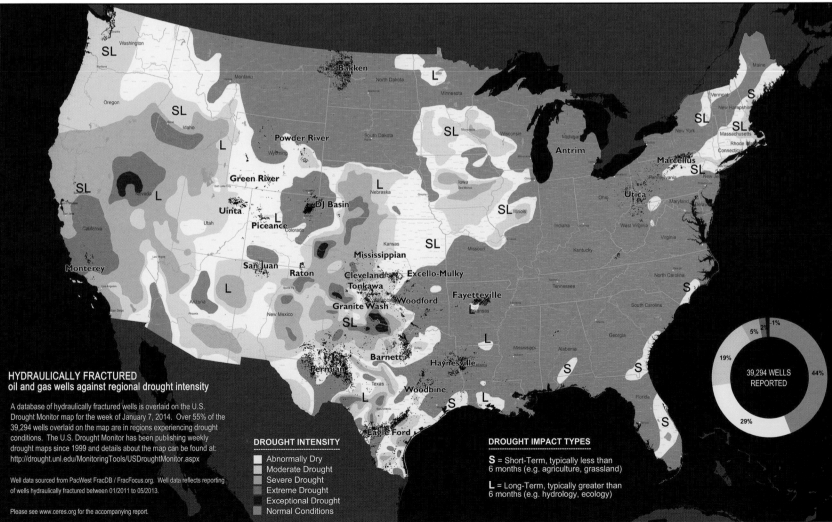

HYDRAULICALLY FRACTURED
oil and gas wells against regional drought intensity

A database of hydraulically fractured wells is overlaid on the U.S. Drought Monitor map for the week of January 7, 2014. Over 55% of the 39,294 wells overlaid on the map are in regions experiencing drought conditions. The U.S. Drought Monitor has been publishing weekly drought maps since 1999 and details about the map can be found at: http://drought.unl.edu/MonitoringTools/USDroughtMonitor.aspx

Well data sourced from PacWest FracDB / FracFocus.org. Well data reflects reporting of wells hydraulically fractured between 01/2011 to 05/2013.

Please see www.ceres.org for the accompanying report.

DROUGHT INTENSITY
- Abnormally Dry
- Moderate Drought
- Severe Drought
- Extreme Drought
- Exceptional Drought
- Normal Conditions

DROUGHT IMPACT TYPES

S = Short-Term, typically less than 6 months (e.g. agriculture, grassland)

L = Long-Term, typically greater than 6 months (e.g. hydrology, ecology)

39,294 WELLS REPORTED — 44% / 29% / 19% / 5% / 2% / 1%

The State of São Paulo Waterways Map

Engemap Engenharia, Mapeamento e Aerolevantamento, Ltda (Engemap Engineering, Mapping, and Aerial Survey, Ltd.)

São Paulo, São Paulo, Brazil

By Engemap Engineering, Mapping, and Aerial Survey, Ltd.

Contact

Weber Pires

weber@engemap.com.br

Software

ArcGIS 10.1 for Desktop

Data Source

Departamento Hidroviario geodatabase version January 2014

The *State of São Paulo Waterways Map* shows the interconnected waterway transport to road and rail loops. The Tietê-Paraná Waterway integrates a large multimodal transport system, presenting itself as an alternative export corridor covering the states of São Paulo, Paraná, Mato Grosso do Sul, Goiás and Minas Gerais. This region of 760,000 square kilometers (293,000 square miles) generates almost half of the Brazilian domestic product.

Courtesy of Engemap Engineering, Mapping, and Aerial Survey, Ltd.

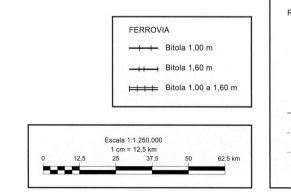

FERROVIA

——+—— Bitola 1,00 m

——++—— Bitola 1,60 m

—++++— Bitola 1,00 a 1,60 m

Escala 1:1.250.000
1 cm = 12,5 km
0   12,5   25   37,5   50   62,5 km

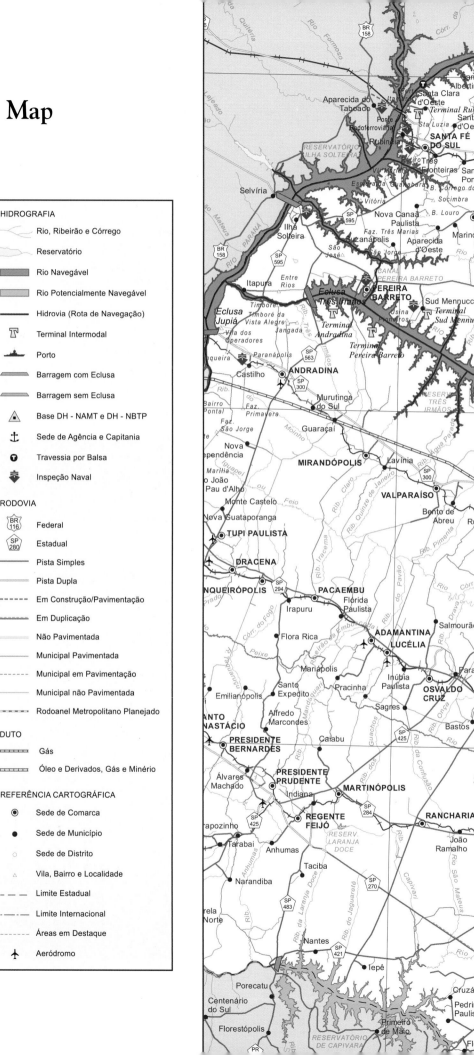

HIDROGRAFIA

Rio, Ribeirão e Córrego

Reservatório

Rio Navegável

Rio Potencialmente Navegável

Hidrovia (Rota de Navegação)

Terminal Intermodal

Porto

Barragem com Eclusa

Barragem sem Eclusa

Base DH - NAMT e DH - NBTP

Sede de Agência e Capitania

Travessia por Balsa

Inspeção Naval

RODOVIA

BR 116 Federal

SP 280 Estadual

Pista Simples

Pista Dupla

Em Construção/Pavimentação

Em Duplicação

Não Pavimentada

Municipal Pavimentada

Municipal em Pavimentação

Municipal não Pavimentada

Rodoanel Metropolitano Planejado

DUTO

Gás

Óleo e Derivados, Gás e Minério

REFERÊNCIA CARTOGRÁFICA

Sede de Comarca

Sede de Município

Sede de Distrito

Vila, Bairro e Localidade

Limite Estadual

Limite Internacional

Áreas em Destaque

Aeródromo

Water Source Protection Zones of Salt Lake County

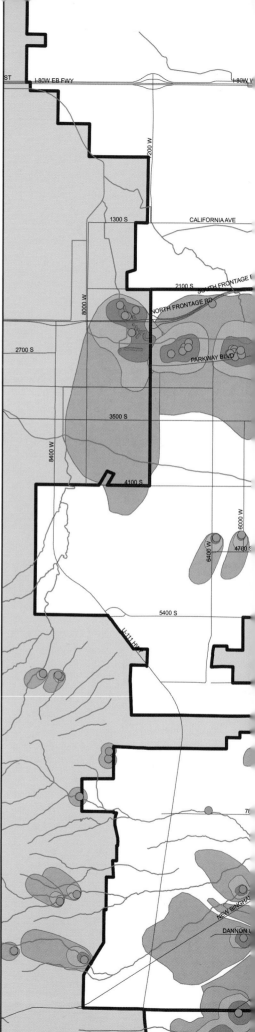

Salt Lake County

Salt Lake City, Utah, USA
By Thomas C. Zumbado

Contact
Thomas C. Zumbado
tzumbado@slco.org

Software
ArcGIS 10.1 for Desktop

Data Source
Salt Lake County Planning and Development Services

In Salt Lake County, where the nearest body of water is an acrid lake with salinity higher than the ocean, groundwater sources are a closely guarded treasure. The construction of homes, commercial areas, and infrastructure must pass close scrutiny on how it affects groundwater areas before any dirt is moved. This map is a visualization of the groundwater GIS data used to reference and vet all land-use development in the county. A combination of well heads, naturally occurring springs, and the plume-like subterranean percolation zones indicate where the groundwater is collected and flows on its way to its respective extraction point. The polygonal percolation zones are divided by the length of time it takes for a drop of water to flow to the well or spring. This duration is known as the TOT, or Time of Travel. When the Planning Department staff receives a prospective land-use project, the corresponding land parcel is cross-referenced with this map to see if the potential to disturb groundwater exists. If so, the applicant can take a number of steps to ensure the safety and integrity of the life-giving water beneath the valley.

Courtesy of Salt Lake County Planning and Development Services.

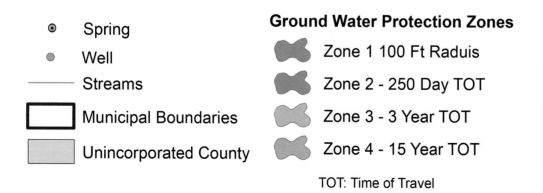

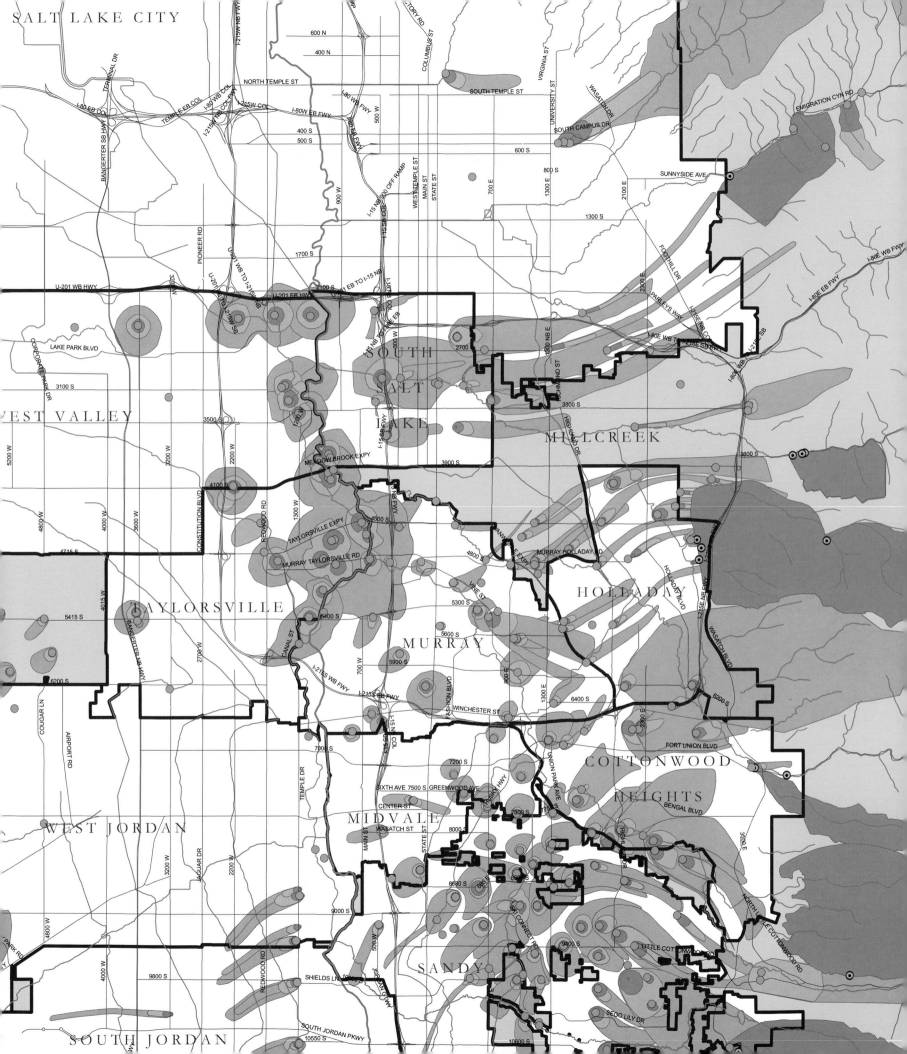

Hydrogeologic Framework and Computer Simulation of Groundwater Flow

New Jersey Department of Environmental Protection

Trenton, New Jersey, USA
By Zehdreh Allen-Lafayette and Laura J. Nicholson

Contact
Zehdreh Allen-Lafayette
zehdreh.allen-lafayette@dep.nj.gov

Software
ArcGIS for Desktop; Adobe Illustrator, Photoshop and InDesign; Microsoft Word, and Corel WordPerfect

Data Sources
New Jersey Department of Environmental Protection, New Jersey Geological and Water Survey

These maps were part of a study in response to a projected increase in groundwater demand in the largely rural Germany Flats area of Sussex County in northern New Jersey. The study by Laura J. Nicholson, "Hydrogeologic Framework and Computer Simulation of Groundwater Flow in the Valley-Fill and Fractured-Rock Aquifers," assessed the hydrogeology and groundwater resource potential of the area.

Courtesy of New Jersey Department of Environmental Protection.

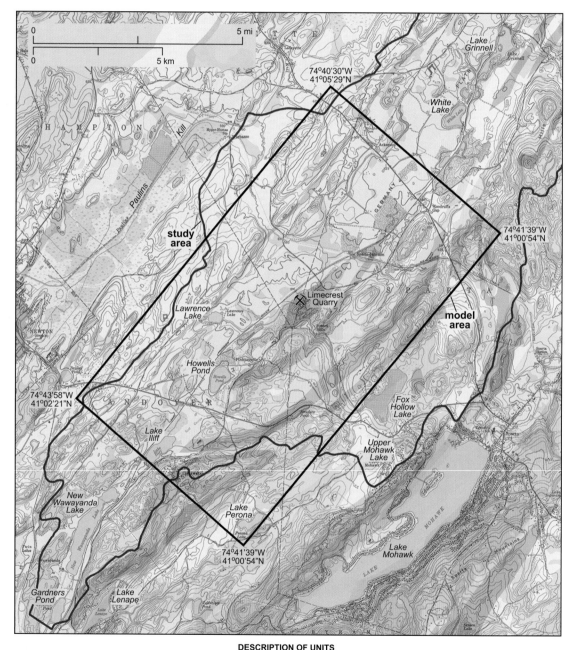

DESCRIPTION OF UNITS

Omr	Ramseyburg Member (Upper & Middle Ordovician)
Omb	Martinsburg Formation, undivided (Upper & Middle Ordovician)
Ojw	Jacksonburg Limestone & Sequence at Wantage (Upper & Middle Ordovician)
OCb	Beekmantown Group (Lower Ordovician & Upper Cambrian)

Ca	Allentown Dolomite (Upper Cambrian)
Chl	Leithsville Formation & Hardyston Quartzite (Middle to Lower Cambrian)
Ybu	Byram Intrusive Suite, undivided (Mesoproterozoic)
Ylhu	Lake Hopatcong Instrusive Suite, undivided (Mesoproterozoic)

Yf	Marble (Mesoproterozoic)
Ymu	Metasedimentary & metavolcanic rocks, undivided (Mesoproterozoic)
Ylu	Losee Metamorphic Suite, undivided (Mesoproterozoic)
Yu	Amphibolite, mafic gneiss & microantiperthite alaskite, undivided (Mesoproterozoic)

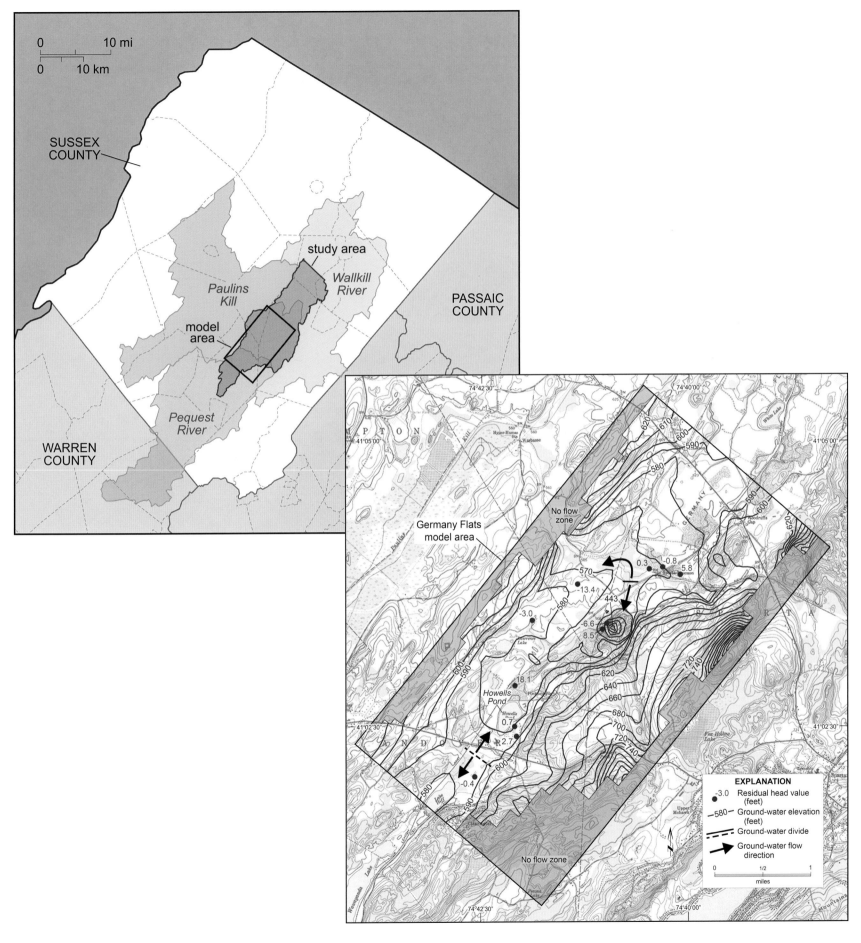

Prioritization for Targeted Watershed Improvement

Furman University

Greenville, South Carolina, USA
By Suresh Muthukrishnan

Contact
Suresh Muthukrishnan
suresh.muthukrishnan@furman.edu

Software
ArcGIS 10.2 for Desktop

Data Sources
US Geological Survey, Greenville County, Esri, Upstate Forever, Greenville; US Environmental Protection Agency; Furman University

The population of Greenville County, South Carolina, has increased by 200 percent since the 1950s. Urban sprawl and other land-cover transformations have put stress on the river ecosystems resulting in increased stream bank erosion, channel widening, channel incision, and floodplain abandonment. This study identifies areas with high potential for flooding, stream channel erosion, and floodplain abandonment. For this study, the Upper Reedy River Watershed was subdivided into 157 smaller subwatersheds, ranging in area from 0.07 to 2.84 square kilometers (0.027 to 1.09 square miles), using an automated watershed delineation method in GIS. The results rank subwatersheds according to level of urban impact and the potential for implementing storm water reduction programs such as rainwater harvesting. Further ground-based study of the areas will help local governments take appropriate action to alleviate river-related problems.

Courtesy of Furman University.

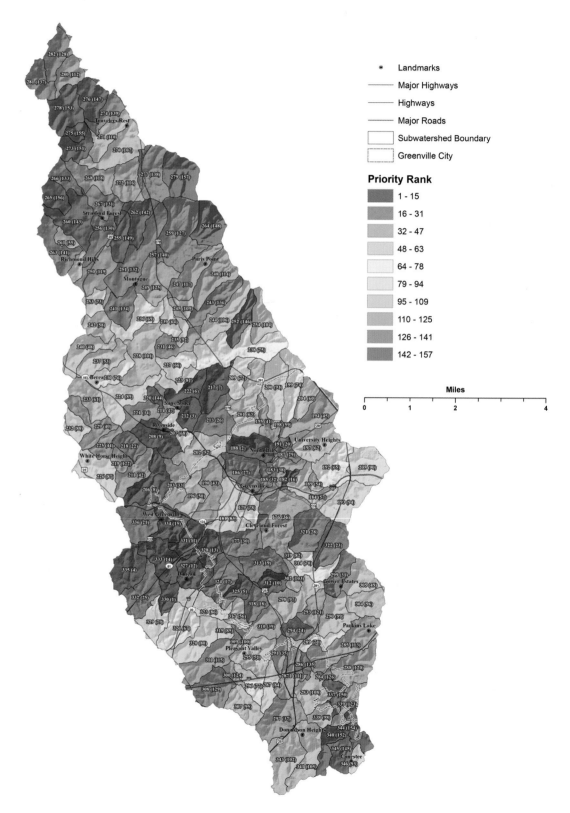

Groundwater Vulnerability Mapping in the Ozarks

The Nature Conservancy (TNC)

Fayetteville, Arkansas, USA
By Cory Gallipeau

Contact
Cory Gallipeau
cgallipeau@tnc.org

Software
ArcGIS for Desktop

Data Sources
US Geological Survey, Natural Resources Conservation Service, Oregon State University, Arkansas Water Resource Center

Groundwater vulnerability mapping helps The Nature Conservancy (TNC) assess areas most susceptible to groundwater contamination. The groundwater vulnerability map here provides a first attempt at developing an Arkansas Ozark karst habitat map. Karst topography means that the geologic structures underneath the earth are made of soluble rock such as limestone and dolomite. The Ozarks are a karst landscape made up of numerous caves, sinkholes, springs, and bluffs. These features hold many rare, unique, threatened, and endangered species. Often, these species have adapted to survive the rigorous environmental conditions in subterranean environments, including a lack of light and limited food. Many of these organisms rely on groundwater resources for their survival. Knowing how vulnerable the landscape is to groundwater infiltration helps TNC focus efforts to protect these natural features and preserve the life that depends on them.

Courtesy of The Nature Conservancy.

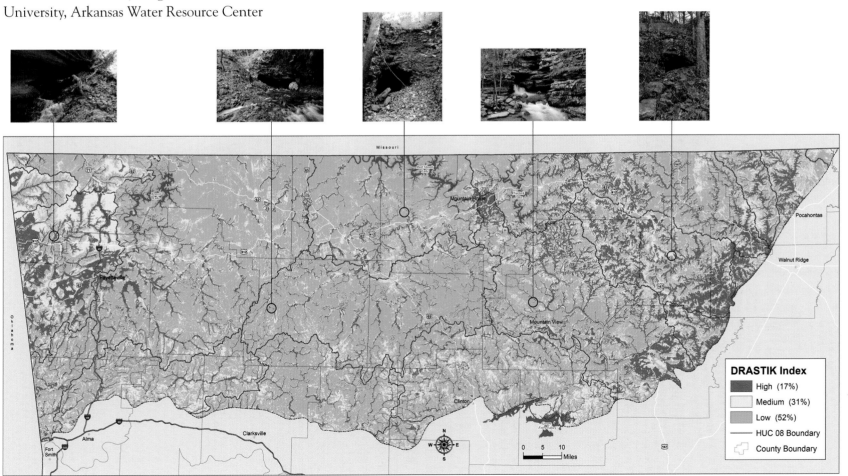

Water Quality Monitoring in the Southeastern United States since 2010

US Environmental Protection Agency (EPA)

Atlanta, Georgia, USA
By Jon Becker

Contact
Jon Becker
becker.jon@epa.gov

Software
ArcGIS 10.1 for Desktop

Data Sources
US Environmental Protection Agency, US Geological Survey, Esri

This map summarizes data on the physical and chemical water quality monitoring in eight southeastern US states between 2010 and early 2014. The US Environmental Protection Agency (EPA) queried sampling results from the national Water Quality Portal and totaled the number of discrete records for each monitoring site. A kernel density analysis created a heat map of monitoring activity based on both the density of the sites and the number of records for each site. Stations and results were also summarized by hydrologic unit boundary watersheds to help identify areas with the most monitoring activity and areas that may have monitoring gaps. While the vast majority of the sampling events reflected here are from surface water, some of them are from groundwater, sediment, air, or other media categories. This map does not capture all of the monitoring being performed in the Southeast, just the records reflected in the US Geological Survey's National Water Information System and the EPA's STOrage and RETrieval (STORET) database. STORET is a repository for water quality data collected by state environmental agencies, various federal agencies, tribes, and others.

Courtesy of Jon Becker, US Environmental Protection Agency.

HUC8s – Watershed Boundary Dataset

USGS NWIS Monitoring Sites
Discrete Records Per Site

- 1 - 100
- 101 - 200
- 201 - 400
- 401 - 700
- 701 - 1,000
- 1,001 - 2,000
- 2,001 - 4,000
- 4,001 - 10,660

STORET Monitoring Sites
Discrete Records Per Site

- 1 - 100
- 101 - 200
- 201 - 400
- 401 - 700
- 701 - 1,000
- 1,001 - 2,000
- 2,001 - 4,000
- 4,001 - 10,660

USA States (below 1:3m)

Rivers and Streams

Water bodies

Kernel Density Gradient Based on Site Density and Number of Records Per Site

- 0 - 0.64
- 0.65 - 2.8
- 2.9 - 5
- 5.1 - 7.2
- 7.3 - 9.3
- 9.4 - 12
- 13 - 14
- 15 - 16
- 17 - 18
- 19 - 20
- 21 - 190

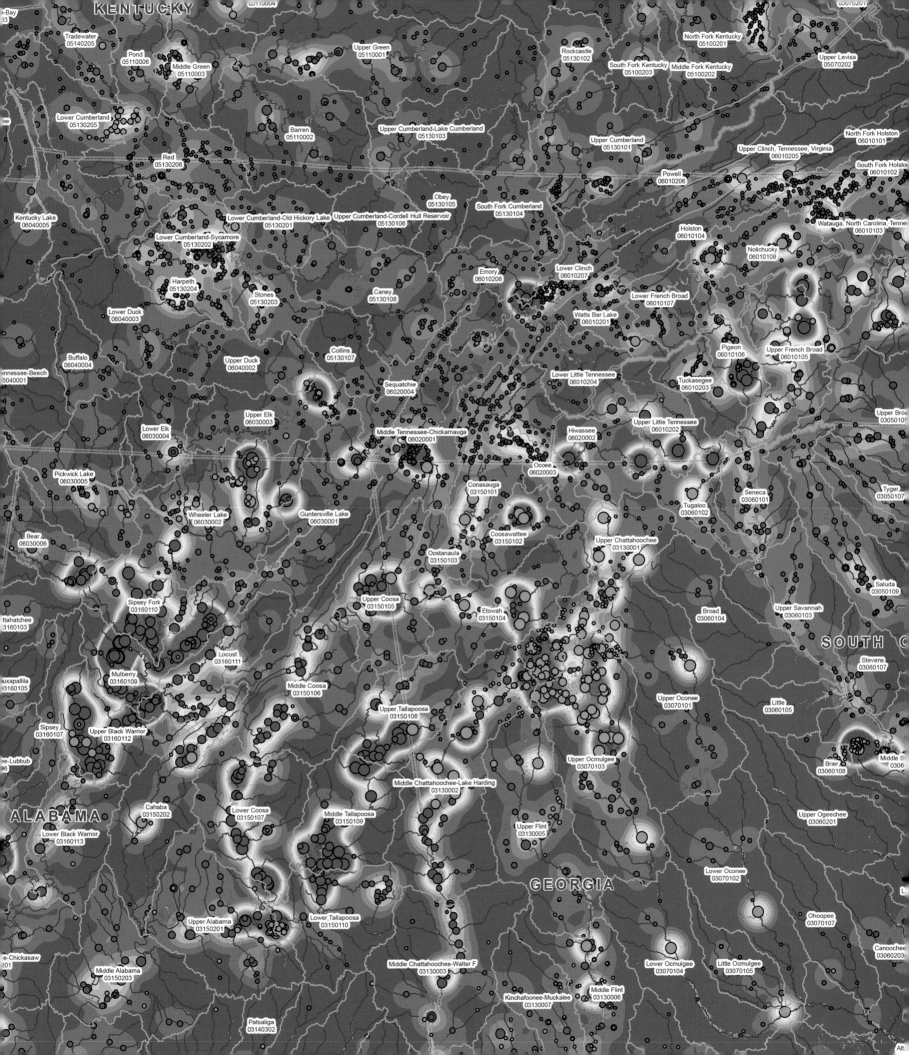

US Army Corps of Engineers Hydrographic Survey Data Management

US Army Corps of Engineers (USACE)

Norfolk, Virginia, USA
By Jeff Swallow

Contact
Jeff Swallow
jeffrey.a.swallow@usace.army.mil

Software
ArcGIS 10.1 for Desktop, Python 2.7

Data Source
US Army Corps of Engineers

The US Army Corps of Engineers (USACE) developed the eHydro application for producing enterprise hydrographic data. This application provides a consistent data management practice for processing, storing, and disseminating hydrographic data to the federal stakeholders of the USACE, including the National Oceanic and Atmospheric Administration and US Coast Guard. Federal regulations define government roles pertaining to navigation data products. USACE Districts are responsible for preparing channel survey/condition maps, data, and reports from the results of each controlled survey of their authorized federal navigation projects.

Courtesy of US Army Corps of Engineers, Norfolk District—Operations Branch; US Army Corps of Engineers, Portland District.

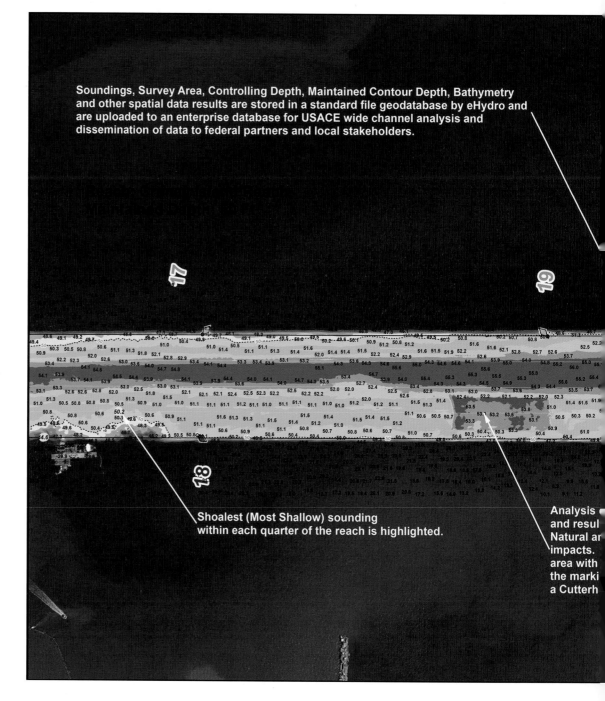

Soundings, Survey Area, Controlling Depth, Maintained Contour Depth, Bathymetry and other spatial data results are stored in a standard file geodatabase by eHydro and are uploaded to an enterprise database for USACE wide channel analysis and dissemination of data to federal partners and local stakeholders.

Shoalest (Most Shallow) sounding within each quarter of the reach is highlighted.

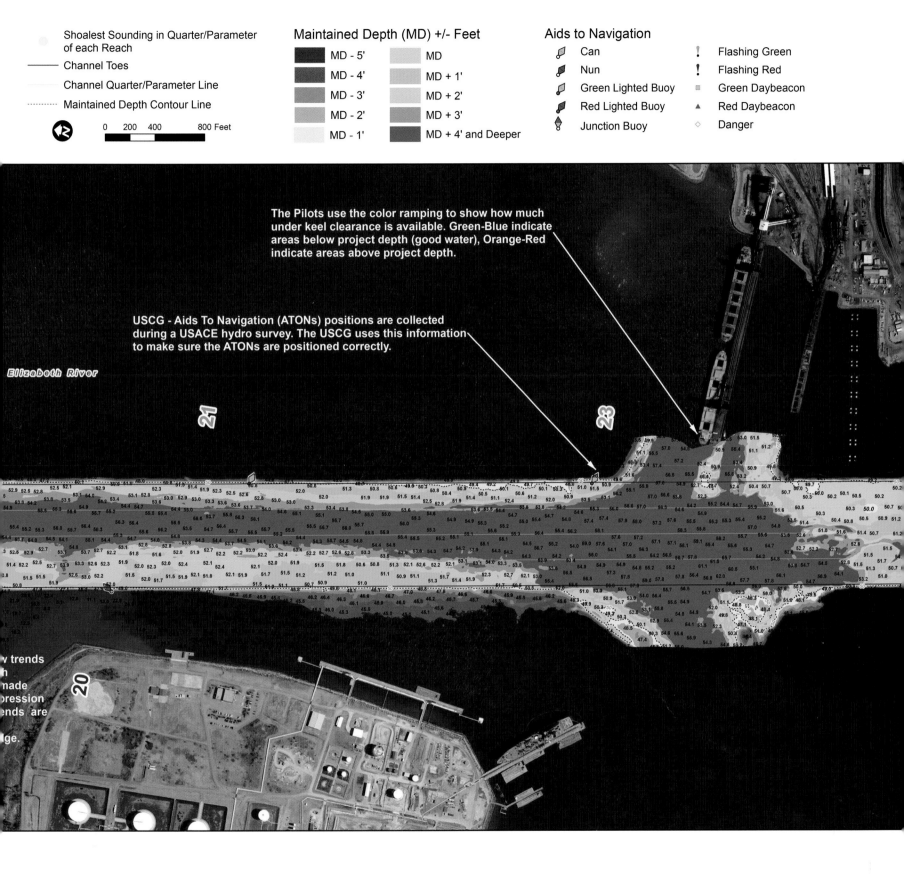

Legend

Shoalest Sounding in Quarter/Parameter of each Reach

— Channel Toes

— Channel Quarter/Parameter Line

······ Maintained Depth Contour Line

0   200   400   800 Feet

Maintained Depth (MD) +/- Feet

MD - 5'	MD
MD - 4'	MD + 1'
MD - 3'	MD + 2'
MD - 2'	MD + 3'
MD - 1'	MD + 4' and Deeper

Aids to Navigation

Can		!	Flashing Green
Nun		!	Flashing Red
Green Lighted Buoy		■	Green Daybeacon
Red Lighted Buoy		▲	Red Daybeacon
Junction Buoy		◇	Danger

The Pilots use the color ramping to show how much under keel clearance is available. Green-Blue indicate areas below project depth (good water), Orange-Red indicate areas above project depth.

USCG - Aids To Navigation (ATONs) positions are collected during a USACE hydro survey. The USCG uses this information to make sure the ATONs are positioned correctly.

Elizabeth River

Measuring Smart Growth Performance and Potential in Rhode Island

Roger Williams University

Bristol, Rhode Island, USA
By Edgar G. Adams Jr. and Brian Boisvert

Contact
Edgar G. Adams Jr.
eadams@rwu.edu

Software
ArcGIS 10.0 for Desktop

Data Source
Rhode Island Geographic Information System

This study helped rural communities with limited planning capacity to quickly evaluate the smart growth potential and development capacity of their designated growth centers. Funding for low and moderate income (LMI) housing is limited to areas within the Urban Services Boundary or designated growth centers. This is a critical factor for ensuring that adequate capacity exists to provide that state-mandated housing. The rural communities in the western portion of Rhode Island are the fastest growing and have the least LMI housing counting toward their 10 percent fair-share goal.

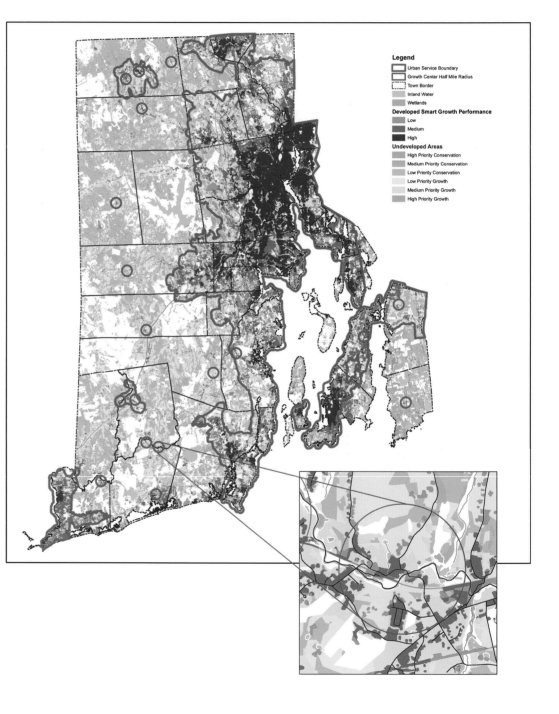

This study was funded by the Roger Williams University Foundation for Teaching and Scholarship. A special thanks to advisory committee members Elizabeth Debs of The Housing Network, Sheila Brush of Grow Smart Rhode Island, and Nancy Hess of Statewide Planning. Brian Boisvert was invaluable as a research assistant for phase I. and Brad Shapiro and Fenton Bradley assisted on phase II, which looked at the location of LMI housing according to a range of related criteria.

Courtesy of Roger Williams University, Edgar G. Adams Jr., and Brian Boisvert.

Got Green Space?

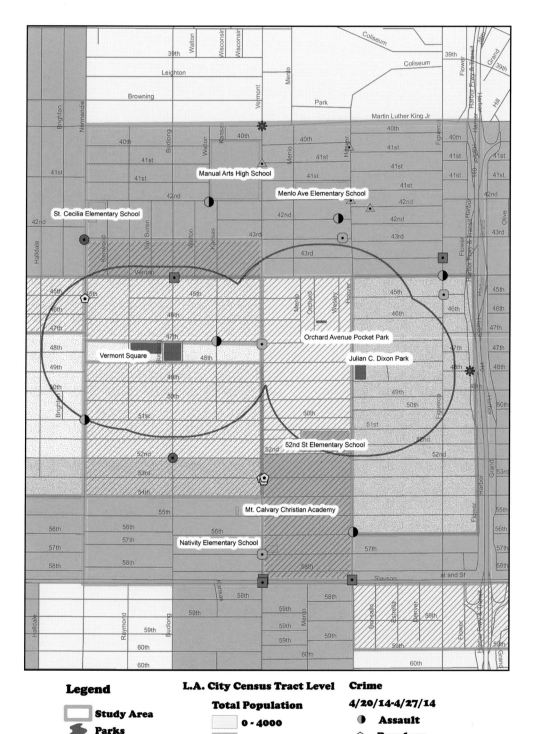

Cal Poly University, Pomona

Pomona, California, USA
By Mary Eleanor Cadena

Contact
Lin Wu
lwu@cpp.edu

Software
ArcGIS 10.1 for Desktop

Data Sources
City of Los Angeles city portal, US Census Bureau, streetgangs.com, crimemapping.com

This student project began with the study of possible benefits of increased green space in the southern area of Los Angeles, where green space is lacking. More green space could mean cleaner air, cleaner water, more recreation areas, and healthier lifestyles. It would also help build a sense of community, encouraging residents to become actively engaged stakeholders for their community. But the site study revealed questionable activities occurring at or near schools and current green spaces, including gun violence. This revelation shifted the focus of the project from green space to safety. The project determined that green areas can improve the quality of life only in safe neighborhoods.

Courtesy of Cal Poly University, Pomona.

Legend

- ▢ Study Area
- ⬧ Parks
- ▮ Schools
- ▦ Crips
- ▨ Sureno
- ▢ 1/2 Mile Walkable Distance to Parks

L.A. City Census Tract Level

Total Population
- 0 - 4000
- 4001 - 8000
- 8001 - 11541

Crime
4/20/14-4/27/14
- ◑ Assault
- ⬠ Burglary
- ✳ Homicide
- △ Larceny
- ◉ Motor Vehicle Theft
- ◨ Robbery
- ▦ Vehicle Break In

Rural Broadband Experiments Funding

James W. Sewall Company

Old Town, Maine, USA
By Randy Claar, Daisy Mueller, and Neal Pickard

Contact
Rick Martens
rmartens@jws.com

Software
ArcGIS 10.1 for Desktop

Data Sources
Eligible Census Tracts: Federal Communications Commission Rural Broadband Experiments, Maine Fiber Company, State Broadband Initiative Mapping Project, ConnectME Authority, 2010 US Census

At the request of the ConnectME Authority and the Maine Office of the Public Advocate, the James W. Sewall Company created a statewide map showing areas eligible to apply for Federal Communications Commission (FCC) Rural Broadband Experiments funding. Along with identifying eligible census tracts, the map also shows housing unit counts in census blocks that do not have broadband-level Internet access.

For the purpose of these experiments, broadband Internet service is defined as a wired or terrestrial fixed wireless connection with maximum advertised speeds of at least 3 Mbps (megabits per second) download and 768 Kbps (kilobits per second) upload. The request was in response to an FCC order to solicit interest from entities willing to deploy robust, scalable broadband networks using Connect America funding. To identify

geographic areas eligible for funding under the Rural Broadband Experiments, the FCC released a list of census tracts. The FCC also stated that preference would be given to applications targeting census blocks that are currently without broadband Internet service.

Courtesy of James W. Sewall Company and ConnectME Authority.

Boundaries

☐ Eligible Census Tracts

▨ Ineligible Census Tracts

☐ Native American Nation Lands

☐ Towns

3 Ring Binder Project

▬▬ Fiber Network

Housing Units in Census Blocks without Broadband

Number of Housing Units: 0

Number of Housing Units: 1 - 15

Number of Housing Units: 16 - 33

Number of Housing Units: 34 - 66

Number of Housing Units: 67 - 180

Community Anchor Institutions

• School-K through 12

• Library

• Medical/healthcare

• Public safety

• University, college, other postsecondary

• Other community support-government

• Other community support-nongovernmental

McNabs and Lawlor Islands Provincial Park

Lost Art Cartography

Grand Pré, Nova Scotia, Canada
By Marcel Morin

Contact
Marcel Morin
cybermapper@gmail.com

Software
ArcGIS Desktop 9.3.1, Adobe Illustrator, Adobe Photoshop

Data Sources
Halifax Regional Municipality Open Data, Friends of McNabs Island Society, SperryDesign

McNabs Island, located at the entrance to Halifax Harbour, is part of McNabs and Lawlor Islands Provincial Park and includes Fort McNab National Historic Site. This 372-hectare (919-acre) natural environment park is only a short boat ride from Halifax, the provincial capital of Nova Scotia. The map was commissioned by the Friends of McNabs Island Society, which has been dedicated to preserving the island as parkland for more than a quarter century. Partnering with Nova Scotia's Department of Natural Resources, the society undertook an inaugural orientation project for island visitors. The orientation map is one component of a welcome kiosk that combines a series of five themed interpretive panels. This map presented distinct challenges. The design had to be attractive and easy to navigate. It had to present island content to interest diverse outdoor enthusiasts, including hikers, picnickers, campers, geocachers, kayakers, birders, and photographers. It also had to be designed for multiple outputs: large, static display, interactive versions for the society's website, and scalable versions for the smartphone app.

Courtesy Marcel Morin Lost Art Cartography.

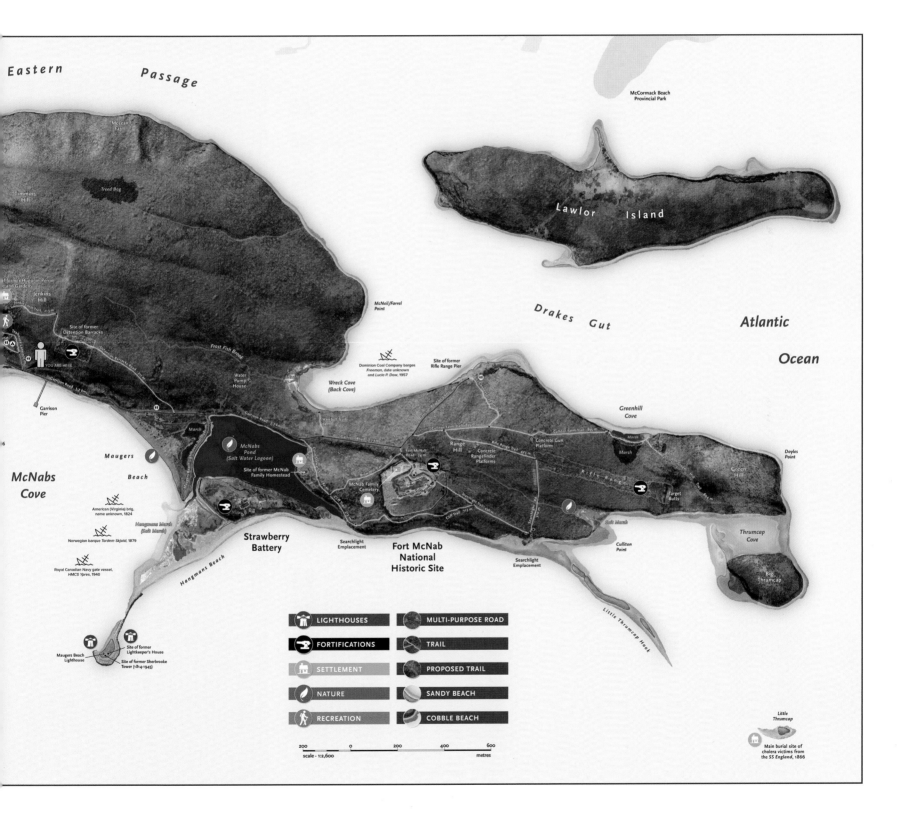

Eastern   Passage

McCormack Beach
Provincial Park

McLean
Farm

Timmins
Hill

Treed Bog

Lawlor   Island

Former Huguenin Ferrin
and Gardens

Jenkins
Hill

McNeil/Farrel
Point

Site of former
Detention Barracks

Drakes   Gut   Atlantic

YOU ARE HERE

Frost Fish Brook

Ocean

Dominion Coal Company barges
Freeman, date unknown
and *Lucia P. Dow*, 1957

Site of former
Rifle Range Pier

Water
Pump
House

Wreck Cove
(Back Cove)

Garrison Pier

Garrison Road 3.7 km

Greenhill
Cove

Marsh

McNabs Pond
(Salt Water Lagoon)

Range
Hill

Concrete Gun
Platform

Marsh

Maugers

Concrete
Rangefinder
Platforms

Fort McNab
Road 74 m

McNabs
Cove

Beach

Site of former McNab
Family Homestead

Rifle Range

Green
Hill

Doyles
Point

American (Virginia) brig,
name unknown, 1824

McNab Family
Cemetery

Target
Butts

Norwegian barque *Tordem Skjold*, 1879

Hangmans Marsh
(Salt Island)

Culliton
Point

Thrumcap
Cove

Royal Canadian Navy gate vessel,
HMCS *Ypres*, 1940

Searchlight
Emplacement

Fort McNab
National
Historic Site

Searchlight
Emplacement

Big
Thrumcap

Strawberry
Battery

Hangmans Beach

Maugers Beach
Lighthouse

Site of former
Lightkeeper's House

Site of former Sherbrooke
Tower (1814-1945)

Little Thrumcap Hook

LIGHTHOUSES		MULTI-PURPOSE ROAD	
FORTIFICATIONS		TRAIL	
SETTLEMENT		PROPOSED TRAIL	
NATURE		SANDY BEACH	
RECREATION		COBBLE BEACH	

200   0   200   400   600

scale - 1:2,600   metres

Little
Thrumcap

Main burial site of
cholera victims from
the SS *England*, 1866

The Tides of Grand Pré

Lost Art Cartography

Grand Pré, Nova Scotia, Canada
By Marcel Morin

Contact
Marcel Morin
cybermapper@gmail.com

Software
ArcGIS Desktop 9.3.1, ArcGIS 3D Analyst, ArcScene, Adobe Illustrator,
Adobe Photoshop

Data Sources
Lost Art Cartography, Municipality of the County of Kings

These views show how the massive Bay of Fundy tides affect the landscape of Grand
Pré, a small historic settlement located in Kings County, Nova Scotia. Grand Pré has a
unique landscape which was recognized by the United Nations Educational, Scientific,
and Cultural Organization (UNESCO) in 2012 as Canada's 16th World Heritage Site.
The Bay of Fundy tides (up to 17 meters or 18 yards vertical range) and evolving changes
to the Minas Basin coastline and river patterns have molded the shape of the Grand Pré
landscape. The dikes have rarely been breached thanks to the continuing efforts of the
Nova Scotia Department of Agriculture. Their retention is critical to the continuance
of agriculture and the inscription of the World Heritage Site. The map series depicts tide
levels and the reduction of land mass in the event of the catastrophic destruction of the
dike walls.

Courtesy of Marcel Morin, Lost Art Cartography.

130

Mid Tide

Mean High Tide

Mean High Tide - Catastrophic dyke wall breach

Minas Basin

Little Island

Boot Island
National Wildlife Area

The Guzzle

North
Grand Pré

ISLAND

LONG

Ransom

Creek

Ransom

Creek

Gaspereau River

Avonport

Creek

Black
Landing

Barn

Creek

Creek

Hortonville

Deluge

Grand Pré

Grand

131

The Landscape of Grand Pré

Lost Art Cartography

Grand Pré, Nova Scotia, Canada
By Marcel Morin

Contact
Marcel Morin
cybermapper@gmail.com

Software
ArcGIS Desktop 9.3.1, Adobe Illustrator, Adobe Photoshop

Data Sources
Lost Art Cartography, Municipality of the County of Kings

Grand Pré is a small historic settlement located in Kings County, Nova Scotia. It has a unique landscape that reflects the use of the salt marsh by aboriginal tribes, followed by the development of dikes and dikeland (1680s to mid-1700s) by Acadian settlers. The United Nations Educational, Scientific, and Cultural Organization (UNESCO) designated Grand Pré a World Heritage Site in 2012. The UNESCO designation is not just related to the area's historical importance but to the continuance of the agricultural land use and cooperative management structure. This map illustrates the fascinating field patterns that continue despite modern agricultural practices. It also shows the extent of the foreshore, which is critical to the development and maintenance of the dikes. The map was commissioned by the Landscape of Grand Pré Society, the collaborative management structure for the World Heritage Site. The map has been proven to be a valuable and fascinating first step in helping visitors and residents understand the unique geography of this UNESCO site.

Courtesy of Marcel Morin, Lost Art Cartography.

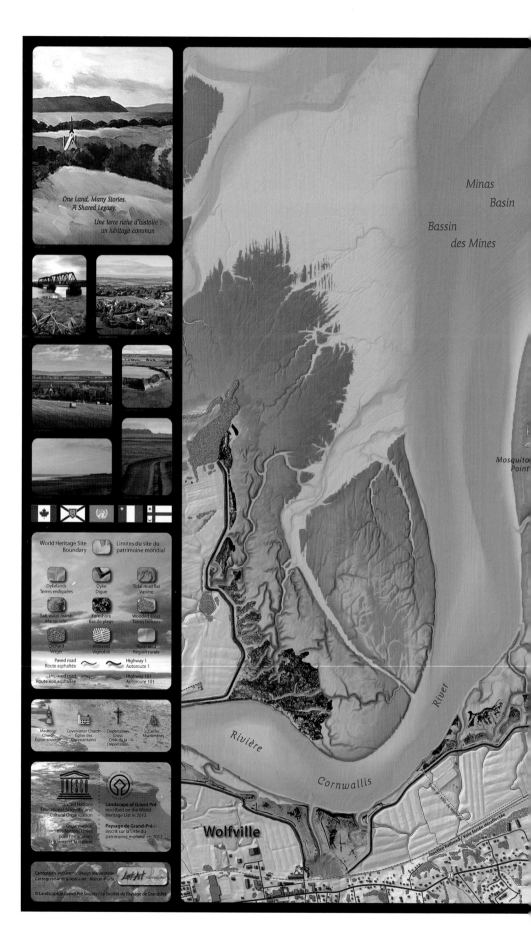

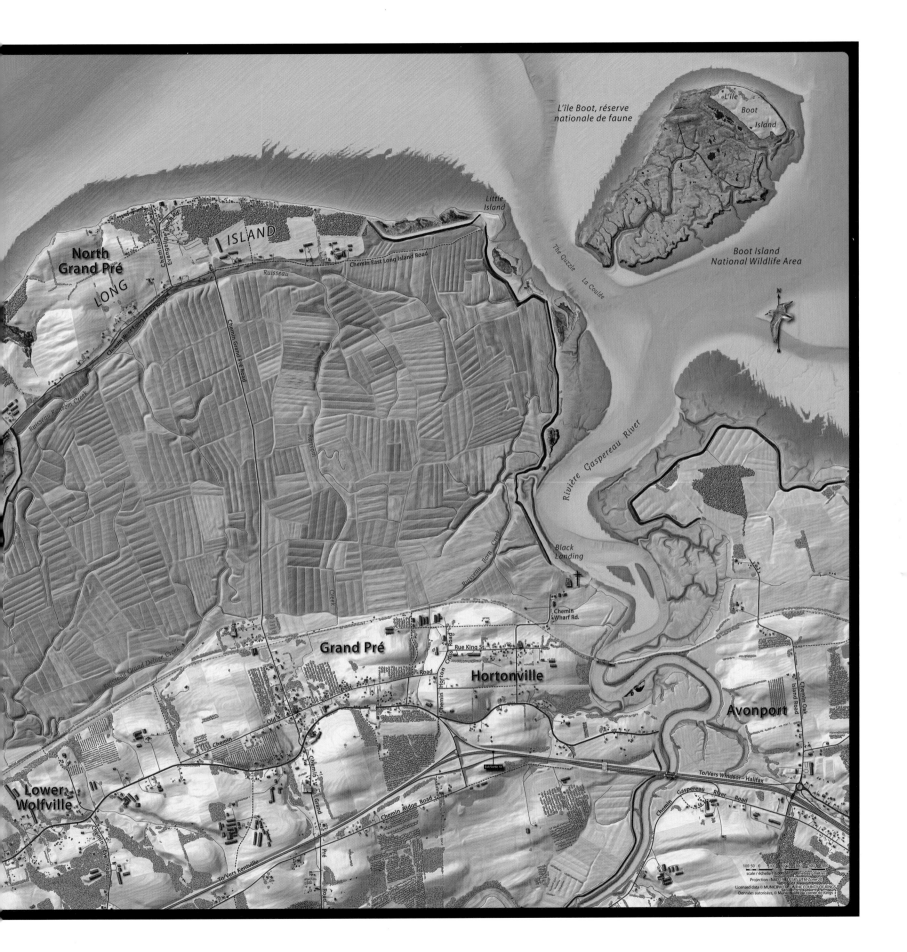

L'île Boot, réserve
nationale de faune

Boot Island
National Wildlife Area

Little
Island

North
Grand Pré

LONG

ISLAND

Chemin East Long Island Road

Ruisseau

Ransom

Creek

Ruisseau Barn Creek

The Guzzle

La Coulée

Rivière Gaspereau River

Black
Landing

Chemin
Wharf Rd.

Grand Pré

Rue King St.

Hortonville

Avonport

Lower
Wolfville

Chemin Ridge Road

To/Vers Windsor – Halifax

Chemin Gaspereau River Road

To/Vers Kentville

Analysis and Visualization of Traffic Data

City of Madison

Madison, Wisconsin, USA
By Dan Seidensticker

Contact

Dan Seidensticker
dseidensticker@cityofmadison.com

Software

ArcGIS 10.2.1 for Desktop, ArcGIS Spatial Analyst, ArcGIS 3D Analyst, ArcScene, Microsoft Excel

Data Sources

Madison Area Transportation Planning Board, National Performance Management Research Data Set, HERE

This analysis is based on the National Performance Management Research Data Set acquired by the Federal Highway Administration which is provided by HERE, a Nokia mapping and location business. Vehicle travel-time data is collected twenty-four hours a day, seven days a week on major highways nationwide from fixed roadside sensors and onboard GPS-enabled devices in vehicles. This data is used for real time traffic maps or for the traffic report on the local

news. In addition to these real time applications, the historical data allows additional opportunities for traffic analysis. For the month of October 2013 in Dane County, Wisconsin, nearly 1 million travel-time recorded events were analyzed with some of the results shown here. The data can be charted in two or three dimensions to visualize traffic congestion, density, or traffic events. The data can also be joined to street centerlines of the National Highway System. The difference between peak and off-peak speeds are calculated for individual road segments and mapped to visualize areas of traffic congestion, as shown here.

Courtesy of Madison Area Transportation Planning Board.

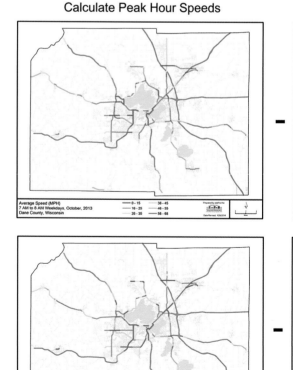

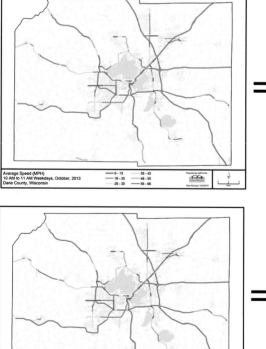

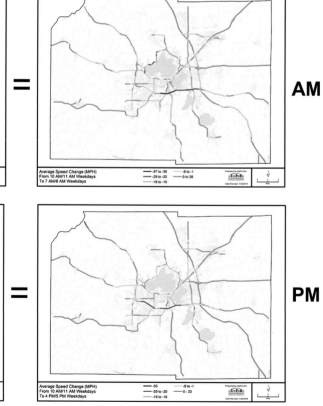

Transport Accessibility of Prague

Palacký University and Charles University

Olomouc and Prague, Czech Republic

By Tomas Hudecek, Zuzana Zakova, Petr Blahnik, Jan Kufner, Radek Churan (analysis authors) and Alena Vondrakova and Vit Vozenilek (map authors)

Contact
Zuzana Zakova
zuz.zakova@seznam.cz

Software
ArcGIS 10.2 for Desktop, Adobe InDesign CS6

Data Source
Research at Charles University, Prague

These maps are part of a project funded by the Grant Agency of the Czech Republic to analyze road- and rail-network development over 100 years, projected to the year 2020. The maps also show Prague's accessibility by road and rail transport. The maps compare the two most frequently used transport modes and illustrate transport accessibility in the country since 1920. A separate data model was created for each period and transport mode. The temporal evaluation of the road network depends on average speed allowed for the particular section. The travel times for the rail network are derived from historic and current timetables.

Courtesy of Palacký University, Olomouc and Charles University, Prague.

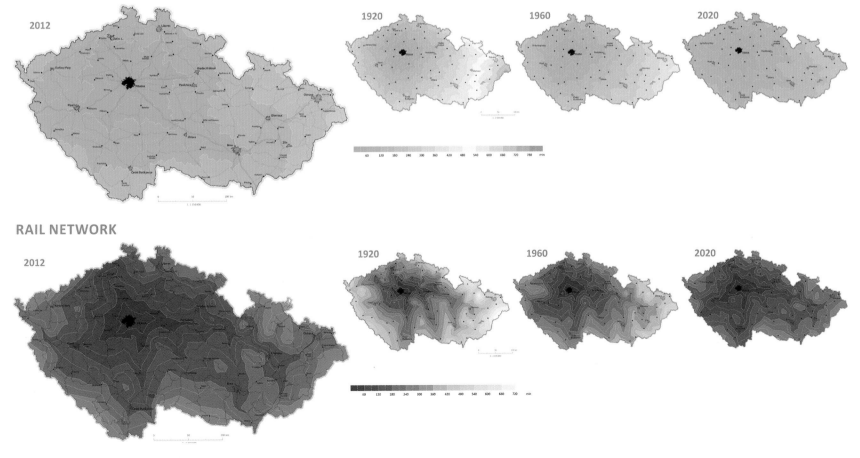

ROAD NETWORK

RAIL NETWORK

TRANSPORTATION

San Diego County Commute-to-Work Analysis

KTU+A

San Diego, California, USA
By Tasha Davis

Contact
Tasha Davis
davis.tasha@gmail.com

Software
ArcGIS 10.1 for Desktop, ArcGIS Network Analyst

Data Sources
San Diego Association of Governments (SANDAG), True North Research,
KTU+A

San Diego County residents were surveyed to assess travel characteristics of their commute. The survey asked for respondents' home and work locations to the nearest cross-street intersection. Using the ArcGIS Network Analyst extension, commute routes along the San Diego County road network were determined for each respondent. The results included the expected busiest road segments based on the shortest route and the resident's likely average commute distance. The information reveals trends in transit travel for the region and facility demand. It also helps transportation planners identify opportunities for network improvement.

Courtesy of KTU+A.

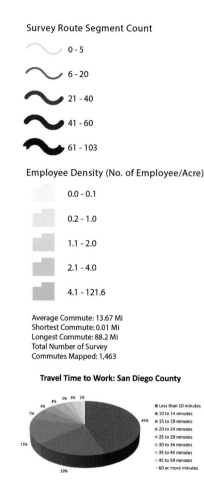

Survey Route Segment Count

0 - 5
6 - 20
21 - 40
41 - 60
61 - 103

Employee Density (No. of Employee/Acre)

0.0 - 0.1
0.2 - 1.0
1.1 - 2.0
2.1 - 4.0
4.1 - 121.6

Average Commute: 13.67 Mi
Shortest Commute: 0.01 Mi
Longest Commute: 88.2 Mi
Total Number of Survey
Commutes Mapped: 1,463

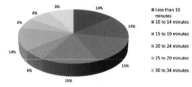

Travel Time to Work: San Diego County

Travel Time to Work: United States

San Diego County
Commute-to-Work

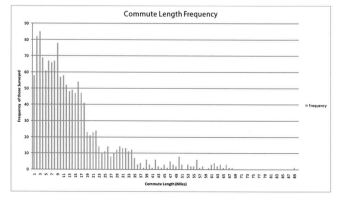

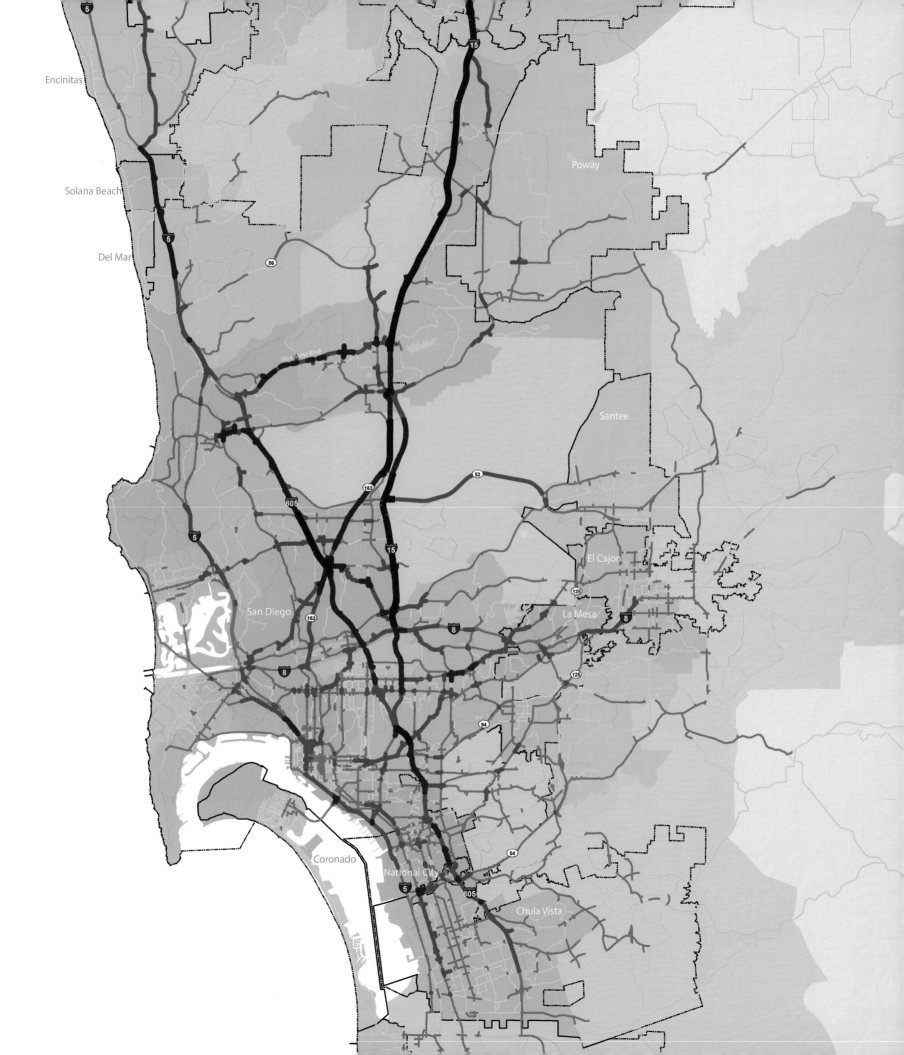

Burbank Bob Hope Airport

City of Burbank

Burbank, California, USA
By Michael A. Carson

Contact
Michael A. Carson
mcarson@burbankca.gov

Software
ArcGIS 10.2 for Desktop

Data Source
City of Burbank

For eighty-four years, Bob Hope Airport in Burbank, California, has been a convenient terminal for travelers flying to or from Los Angeles and the San Fernando Valley. The airport serves over 4 million passengers each year. This map illustrates air traffic data for April 2014. The 7,404 total flights are visualized in different ways, including by arrival and departure, aircraft type, and top arrival/destination airports. Patterns of main flight-line groupings emerge over the Southland. The airport has had many names and uses in its history, Including United Airport in 1930 and Bob Hope Airport since 2003. During World War II, the airport became the main Lockheed production facility for B-17s, Hudson bombers, and P-38 fighters. Later, Lockheed used the airport to work on secret projects that included the U-2 spy plane and the SR-71 Blackbird. Today, the airport is served by major airlines providing nonstop destinations to a dozen major cities.

Courtesy of City of Burbank GIS.

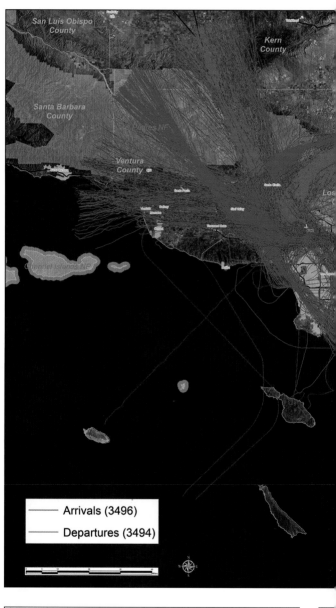

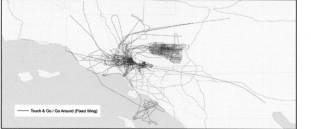

,194)

Regional Jet (456)

Turboprop (809)

Piston (767)

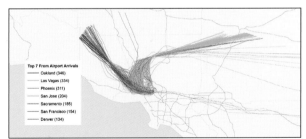

Helicopter (677)
Military (3)

San Bernardino County

Mojave NPRES

Riverside County

Orange County

San Diego County

Imperial County

Top 7 From Airport Arrivals
Oakland (346)
Las Vagas (334)
Phoenix (311)
San Jose (204)
Sacramento (185)
San Francisco (154)
Denver (134)

Top 7 To Airport Departures
Oakland (347)
Las Vagas (329)
Phoenix (309)
San Jose (213)
Sacramento (189)
San Francisco (138)
Denver (121)

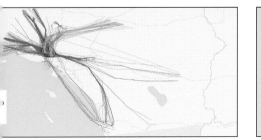

Airlines
Southwest (2456)
SkyWest (793)
Mesa/US Airways (226)
Alaska (174)
SeaPort (105)
JetBlue (60)

Air Operations
Air Carrier (3767)
General Aviation (2872)
Cargo (654)
Commuter (105)
Military (3)
Unknown (3)

Multimodal Map of Brazil—National Plan for Roads

Engemap Engenharia, Mapeamento e Aerolevantamento, Ltda. (Engemap Engineering, Mapping, and Aerial Survey, Ltd.)

São Paulo, São Paulo, Brazil
By Engemap Engineering, Mapping, and Aerial Survey, Ltd.

Contact
Weber Pires
weber@engemap.com.br

Software
ArcGIS 10.1 for Desktop

Data Source
Geodatabase in Oracle and ArcSDE in Departamento Nacional de Infraestrutura de Transportes, Brazil

Engemap produced this map at the request of Brazil's National Department of Transportation Infrastructure (Departamento Nacional de Infraestrutura de Transportes or DNIT). The map shows the multimodal transportation network of Brazil and the network of federal roads, railroads, and waterways. The transportation system of the federal roads network is administered by DNIT under the Brazilian Ministry of Transport.

Courtesy of Engemap Engineering, Mapping, and Aerial Survey, Ltd.

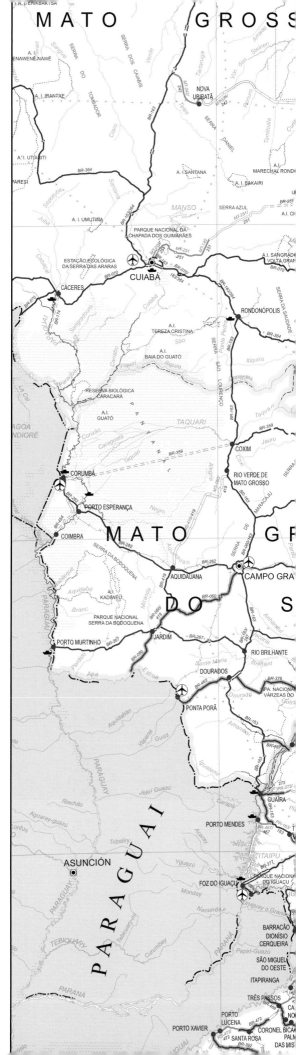

RODOVIAS FEDERAIS
(BR)

Duplicada
Pavimentada
Em Pavimentação
Implantada
Em Implantação
Leito Natural
Planejada
Concedida
Trechos MP 082/2002

RODOVIAS ESTADUAIS COINCIDENTES
(EST/ BR)

Duplicada
Pavimentada
Em Pavimentação
Implantada
Em Implantação
Leito Natural
Planejada
Concedida

LIMITES

Internacional
Interestadual
Interestadual em Litígio — EM LITÍGIO
Parque Nacional, Reserva Florestal e Terras Indígenas

REFERÊNCIAS CARTOGRÁFICAS

Aeródromo Internacional
Porto

HIDROGRAFIA

Rio e Lagoa Permanente
Rio e Lagoa Intermitente

Barragem e Açude
Área Alagadiça

HIDROVIAS

Hidrovia

FERROVIAS

Existente com tráfego/ tráfego suspenso
Em Construção
Planejada

LOCALIDADES

Capital do País
Capital de Estado
Pontos de Passagem

TRANSPORTATION
Aeronautical Map of Suriname

GISsat nv and Civil Aviation Safety Authority Suriname (CASAS)

Paramaribo, Suriname
By GISsat nv

Contact
Stef De Ridder
s.deridder@gissat.com

Software
ArcGIS for Desktop

Data Sources
GISsat nv, CASAS

The Republic of Suriname is located on the northeastern Atlantic coast of South America and is the smallest independent country in South America. This map contains aeronautical information of the main airfields in Suriname and all relevant topographical information for the aviation industry. The map is being reproduced and distributed by CASAS.

Courtesy of GISsat nv, Civil Aviation Safety Authority Suriname.

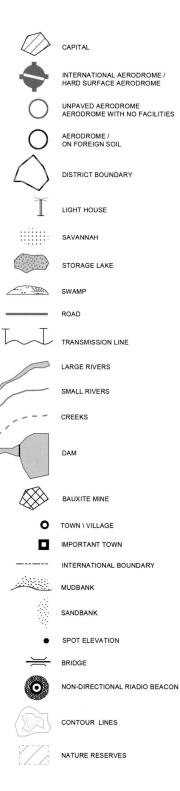

	CAPITAL
	INTERNATIONAL AERODROME / HARD SURFACE AERODROME
	UNPAVED AERODROME AERODROME WITH NO FACILITIES
	AERODROME / ON FOREIGN SOIL
	DISTRICT BOUNDARY
	LIGHT HOUSE
	SAVANNAH
	STORAGE LAKE
	SWAMP
	ROAD
	TRANSMISSION LINE
	LARGE RIVERS
	SMALL RIVERS
	CREEKS
	DAM
	BAUXITE MINE
	TOWN \ VILLAGE
	IMPORTANT TOWN
	INTERNATIONAL BOUNDARY
	MUDBANK
	SANDBANK
	SPOT ELEVATION
	BRIDGE
	NON-DIRECTIONAL RIADIO BEACON
	CONTOUR LINES
	NATURE RESERVES

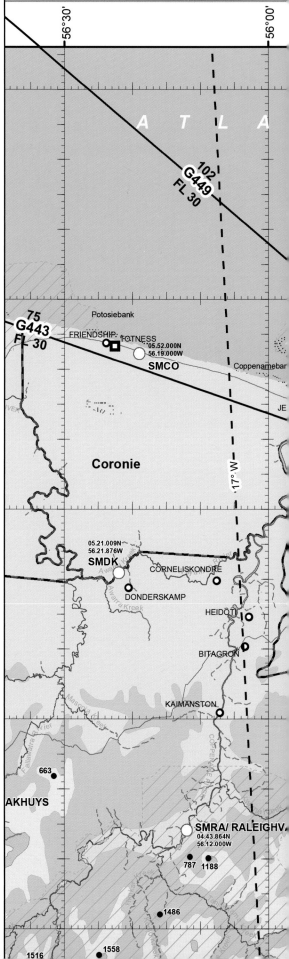

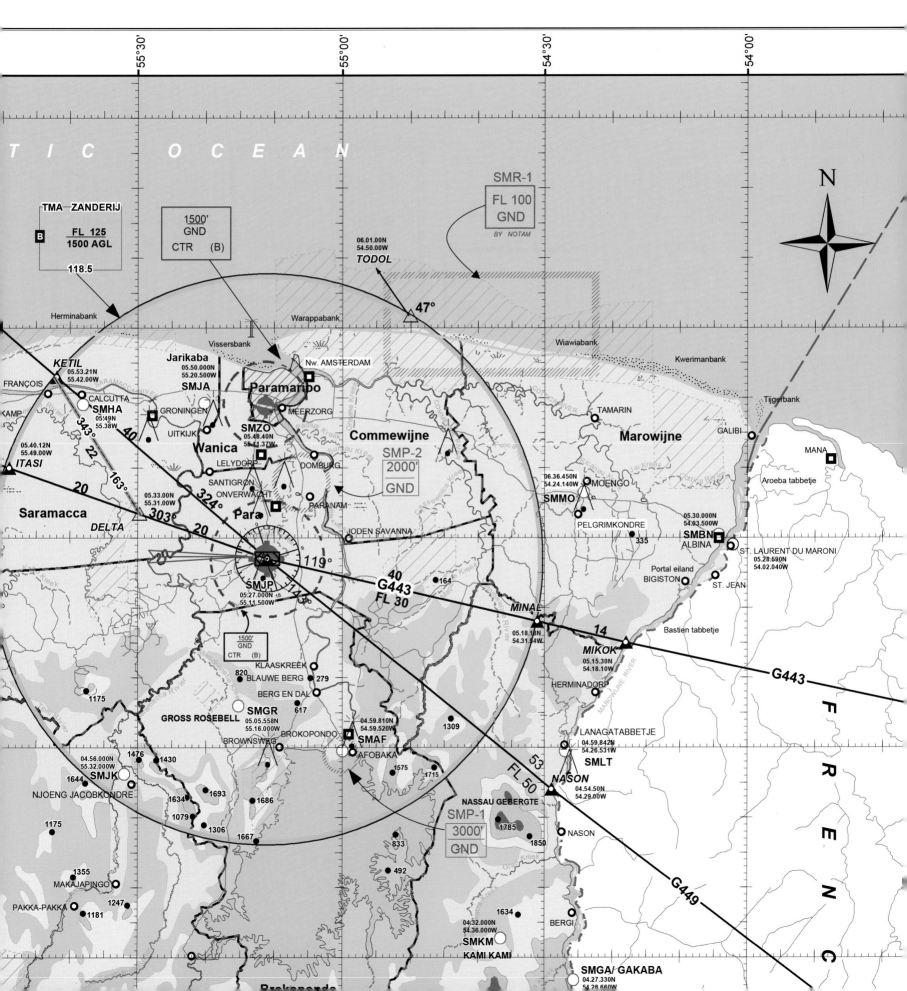

Spatial Analytics for Bicycle/Pedestrian Mobility Planning

KTU+A

San Diego, California, USA
By Kristin Bleile

Contact
Kristin Bleile
kristin@ktua.com

Software
ArcGIS 10.1 for Desktop

Data Sources
San Diego Geographic Information Source, American Community Survey Data, California Highway Patrol, Statewide Integrated Traffic Records System

People want to walk and ride bicycles safely in their neighborhoods. People from a wide variety of backgrounds are forming partnerships with schools, health agencies, neighborhood associations, and environmental organizations to urge their elected officials to create streets and neighborhoods that fit this vision. A comprehensive approach using GIS helps identify bicycle and pedestrian needs. Census data shows high-employment centers and provides statistics for commuting to work by bike or walking. Gaps in pedestrian networks, such as missing sidewalks, hinder pedestrians from walking to destinations such as parks or schools. Bicycle and pedestrian collisions indicate potential problems in the mobility network and provide insight into where safer bicycle or pedestrian conditions should be. These and other factors are ranked in the GIS model. The more favorably ranked segments are then considered for project recommendations.

Courtesy of Kristin Bleile, KTU+A.

Integrated GIS Solution for Smart Grid

Outage Management

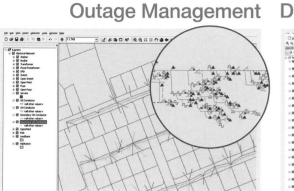

Distribution Management

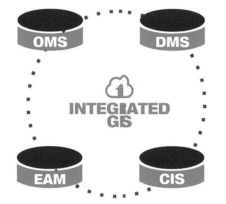

Asset Records Management

Customer Information Management

RMSI

New York, New York, USA
By Amit Rishi

Contact
Amit Rishi
amit.rishi@rmsi.com

Software
ArcGIS for Desktop, Schneider Electric's ArcFM

Data Source
RMSI

A fully integrated GIS enables utilities to have a real-time representation of their assets across various business functions. RMSI implemented a GIS solution with components such as network mapping, asset records management, as-built updates, distribution station modeling, primary and secondary connectivity and integration between customer information systems, and asset management systems. RMSI specializes in providing geospatial data management, engineering design drafting, application development, and consulting services to electric and gas utilities.

Courtesy of RMSI.

145

North American Electric Power System Map

Platts

Denver, Colorado, USA
By Claude Frank and Erin LeFevre

Contact
Claude Frank
claude.frank@platts.com

Software
ArcGIS 10.1 for Desktop, Adobe Illustrator CS6

Data Source
Map Data Pro from Platts

The *North American Electric Power System Map* is the industry standard for understanding the generation, transmission, and utility landscape in the United States and Canada. The map is the cornerstone of Platts' electric power suite of maps. This classic view of the electric power system has been helping businesses understand power for over fifteen years. Updated annually to reflect the latest generation and transmission projects and utility mergers and acquisitions, the map shows hundreds of planned power generation projects, planned transmission, changes in utility service, and comprehensive coverage of the existing infrastructure.

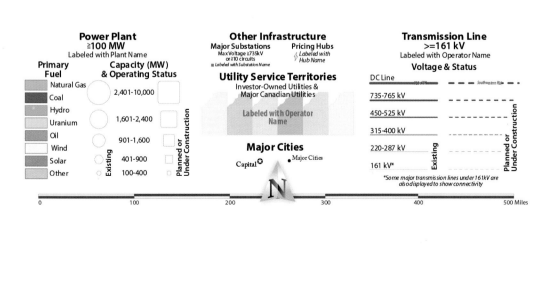

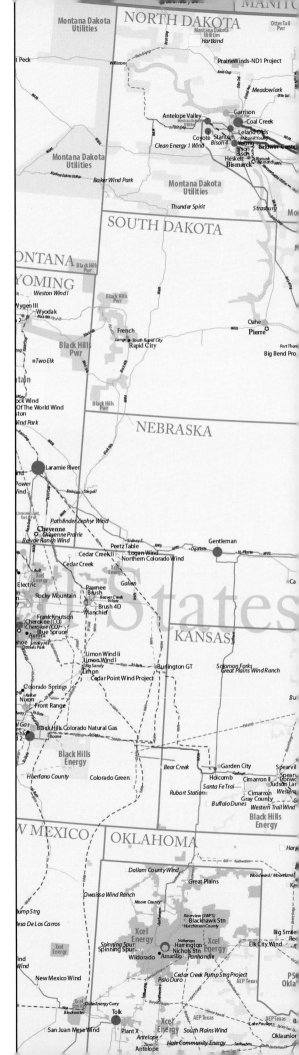

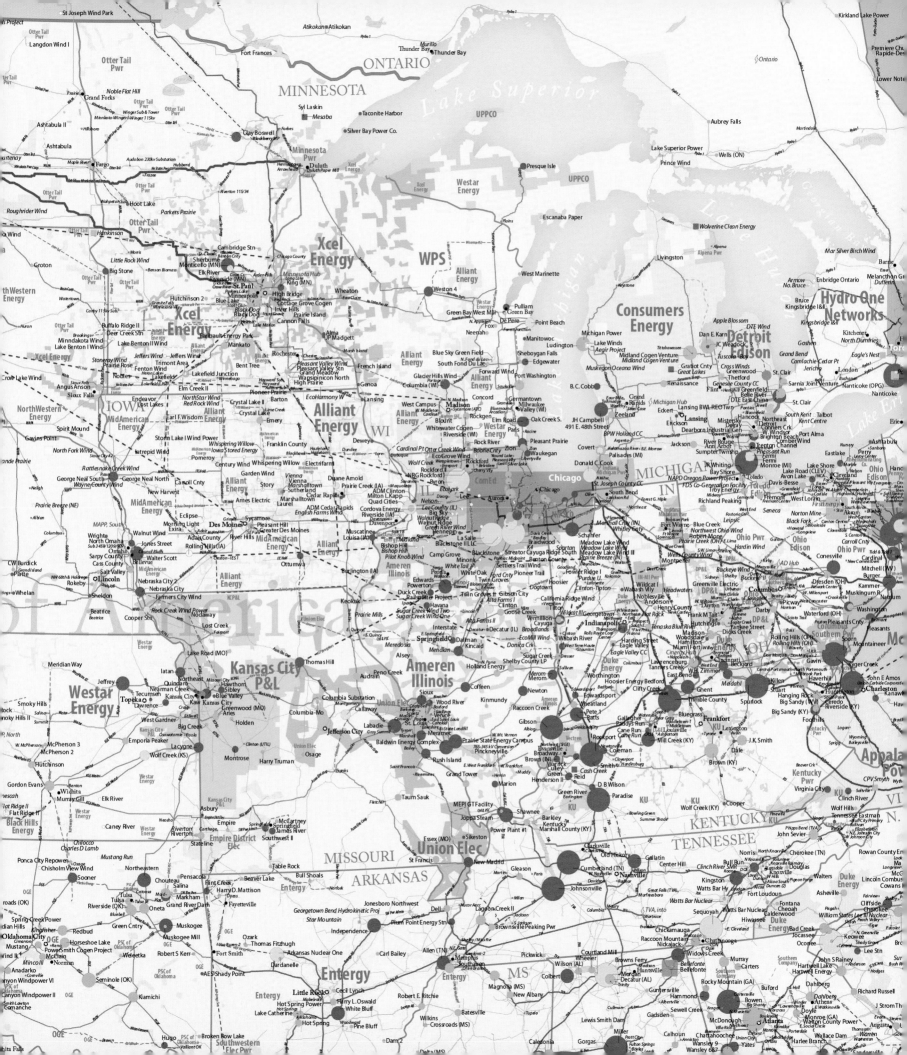

National Interconnected Power System — Brazil

Engemap Engenharia, Mapeamento e Aerolevantamento, Ltda.
(Engemap Engineering, Mapping, and Aerial Survey, Ltd.)

São Paulo, São Paulo, Brazil
By Engemap Engineering, Mapping, and Aerial Survey, Ltd.

Contact
Weber Pires
weber@engemap.com.br

Software
ArcGIS 10.1 for Desktop

Data Source
Geodatabase from the Operador Nacional do Sistema Elétrico Brasileiro, version April 2014

Engemap produced this map at the request of the National System Operator. The map is based on technical information from the geographical distribution of electric transmission networks of the National Interconnected System. The National System Operator is responsible for coordinating and controlling the generation and transmission of electricity in the National Interconnected System under the supervision and regulation of the National Electric Energy Agency.

Courtesy of Engemap Engineering, Mapping, and Aerial Survey, Ltd.

Componentes do Sistema

▲ Usina Hidrelétrica
△ Usina Hidrelétrica Planejada
C Usina Térmica a Carvão
G Usina Térmica a Gás Natural
O Usina Térmica a Óleo Combustível/Diesel
N Usina Térmica Nuclear
B Usina Térmica a Biomassa
P Usina Térmica a Gás de Processo
E Usina Eólica
C Usina Térmica a Carvão Planejada
G Usina Térmica a Gás Natural Planejada
O Usina Térmica a Óleo Combustível/Diesel Planejada
B Usina Térmica a Biomassa Planejada
P Usina Térmica a Gás de Processo Planejada
E Usina Eólica Planejada
● Subestação
○ Subestação Planejada
▶ Conversora
▷ Conversora Planejada

Componentes das LTs

—2— Número de circuitos
----(03)---- Data provável de entrada em operação da instalação.
—FRB— Instalações fora da Rede Básica.
⟁ Futuro seccionamento de LT.

Linhas de Transmissão

Existente	Planejada	Tensão
———	▪▪▪▪▪▪	± 800 kV
———	▪▪▪▪	750 kV
═══	═══	± 600 kV CC
———	▪▪▪▪	500 kV
———	▪▪▪▪	440 kV
———	▪▪▪▪	345 kV
—•—•—		Cabo Subterrâneo
—•—•—		Cabo Subterrâneo
———		230 kV
———		138 kV
———	-----	≤ 88 kV

Referências Cartográficas

◆ Capitais
○ Localidades
– – Limite Internacional
▬▬ Limite Nacional
— — Limite Estadual

Componentes das Usinas

△(03) Data provável de entrada em operação da instalação.

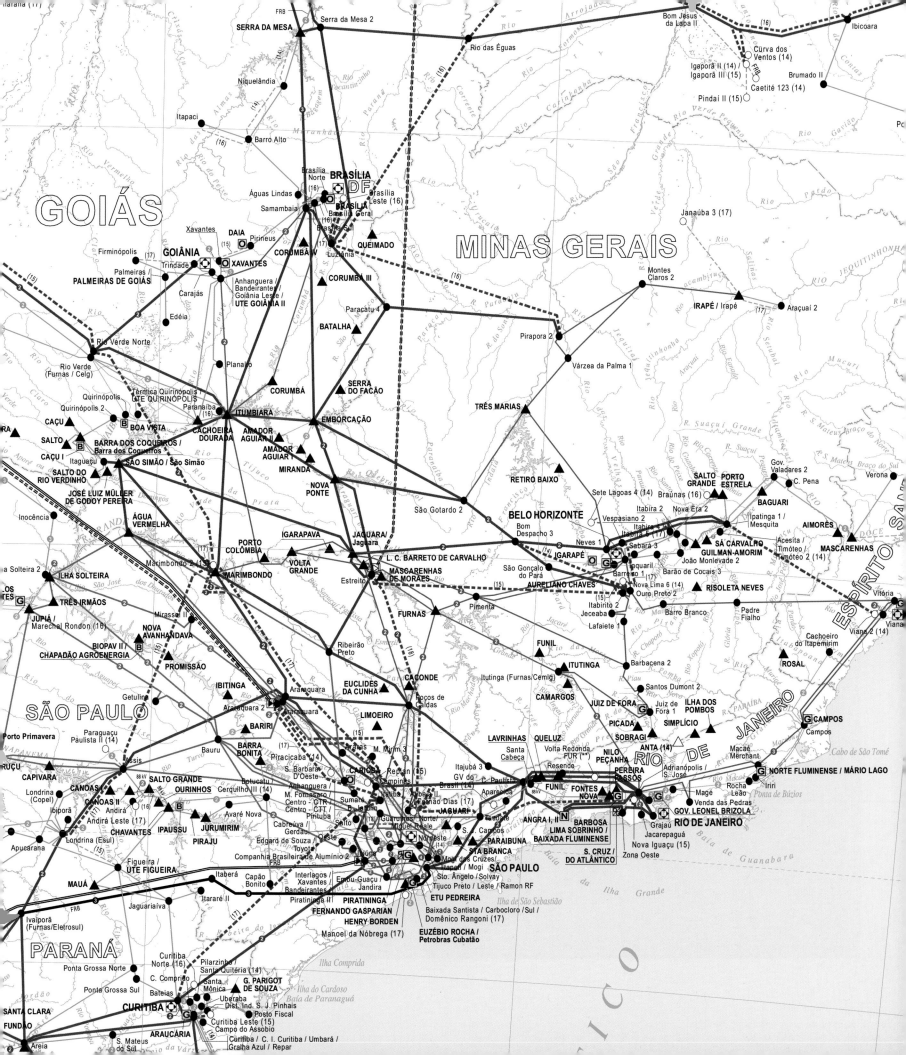

When Superstorms Strike: An Electric Utility's Perspective

National Grid

Syracuse, New York, USA
By John Hayes

Contact
John Hayes
john.hayes@nationalgrid.com

Software
ArcGIS 10.2 for Desktop

Data Sources
National Grid, Iowa State University of Science and Technology

The northeastern United States is no stranger to harsh weather and 2011 was particularly vicious. Several winter storms slammed into mainland Massachusetts. Although winter storms are typical for Massachusetts, the early arrival of snowfall in 2011 left nature in a fragile state. Trees were often still in leaf which, when combined with the added weight of snow and soft ground from the rainy season, increased their risk of uprooting. Massachusetts recorded thirty-two inches of snowfall, resulting in over 400,000 residents without power for days. National Grid prepared for the storm by calling in outside crews and positioning them in strategic locations. The peak of power outages, portrayed in this map, showcases the immense challenge crews faced.

Courtesy of John Hayes, National Grid.

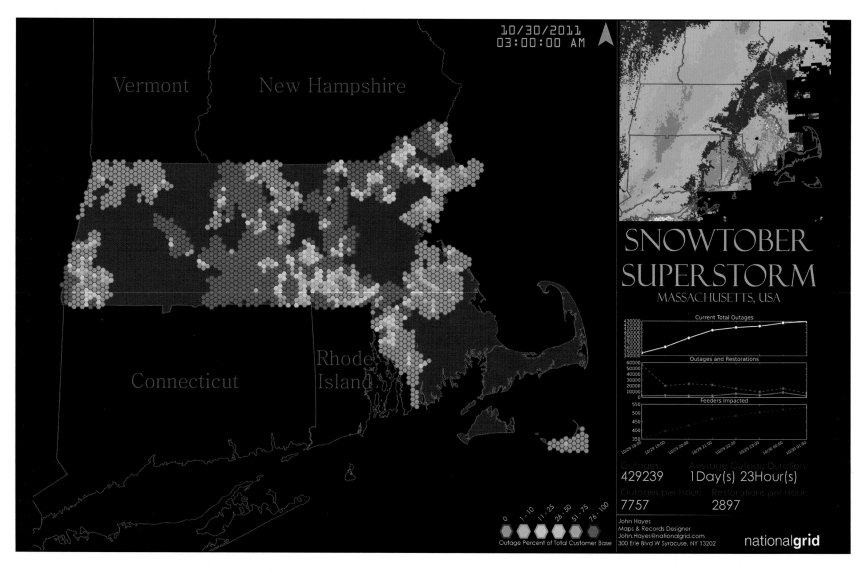

Transformer Capacity Analysis

Murray City

Murray, Utah, USA
By Ben Teran

Contact

Ben Teran
bteran@murray.utah.gov

Software

ArcGIS 10.2 for Desktop

Data Source

Murray City GIS

The analysis presented in this map helped identify all the power system transformers in Murray City that are 140 percent over capacity during peak summer consumption. ArcGIS software was used to join power meter information with customer utility billing information and then trace the flow downstream from each transformer to determine individual transformer load. ArcGIS

ModelBuilder was used to automate that same process for the more than 2,000 transformers within the city. Once the load was calculated for all the transformers, the next step was to identify only those transformers that were at 140 percent over capacity. This process found 106 transformers that the city will consider for replacement to help avoid power outages in the future.

Courtesy of Ben Teran, Murray City GIS.

Developing a California Drinking Water Contaminant Index

Office of Environmental Health Hazard Assessment, California Environmental Protection Agency

Oakland, California, USA
By Komal Bangia, Laura August, Aaron King, John Faust, Andrew Slocombe, Walker Wieland, and George Alexeeff

Contact
Komal Bangia
komal.bangia@oehha.ca.gov

Software
ArcGIS 10.2 for Desktop

Data Sources
California Environmental Health Tracking Program, State Water Resources Control Board, US Census Bureau, US Public Land Survey System

Californians receive their drinking water from a variety of sources and distribution systems. Drinking water quality varies with location, water source, treatment method, and the ability of water providers to remove contaminants before distribution. Drinking water contamination has the potential for widespread effects on health. Although water systems in California have a high rate of compliance with drinking water standards, some low-income and rural communities remain disproportionately exposed to contaminants in their drinking water. To identify areas where contaminated water may be a concern, a drinking water contaminant index was developed by the Office of Environmental Health Hazard Assessment as part CalEnviroScreen—a geographic tool for evaluating multiple pollution sources in a community while accounting for a community's vulnerability to pollution's adverse effects. This drinking water contaminant index compares census tracts across California based on the reported drinking water contaminant concentration data. The index takes into account whether multiple contaminants are present, the measured level of contaminants, and whether the water system has received violations in the past. (Note: Approximated boundaries here are not the same as those used in the CalEnviroScreen analysis. These boundaries and their associated water-quality estimates were altered to protect the confidentiality of drinking water-testing locations and are only presented here to illustrate the methodology.)

Courtesy of Office of Environmental Health Hazard Assessment, California Environmental Protection Agency.

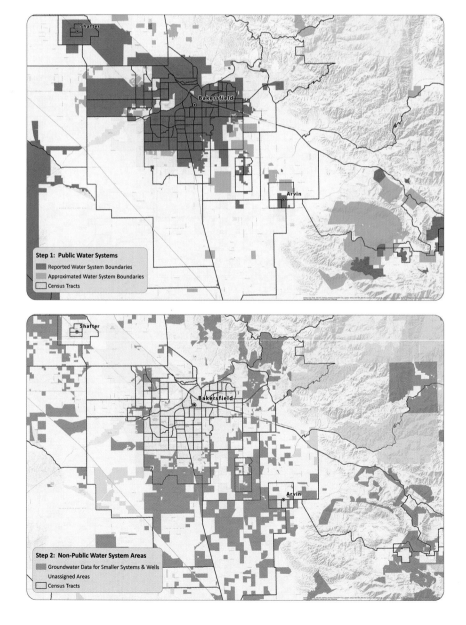

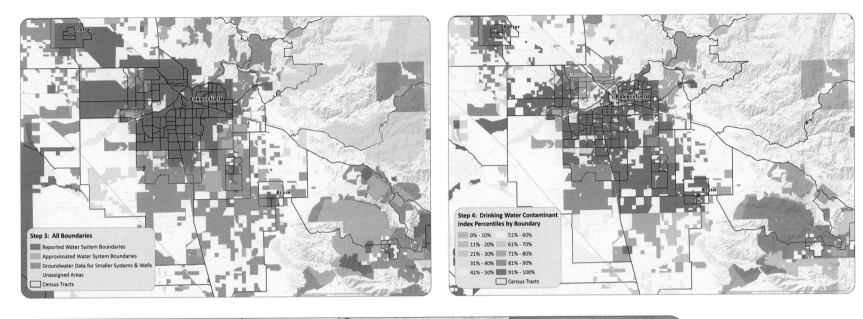

Step 3: All Boundaries

- Reported Water System Boundaries
- Approximated Water System Boundaries
- Groundwater Data for Smaller Systems & Wells
- Unassigned Areas
- Census Tracts

Step 4: Drinking Water Contaminant Index Percentiles by Boundary

0% - 10%	51% - 60%
11% - 20%	61% - 70%
21% - 30%	71% - 80%
31% - 40%	81% - 90%
41% - 50%	91% - 100%

- Census Tracts

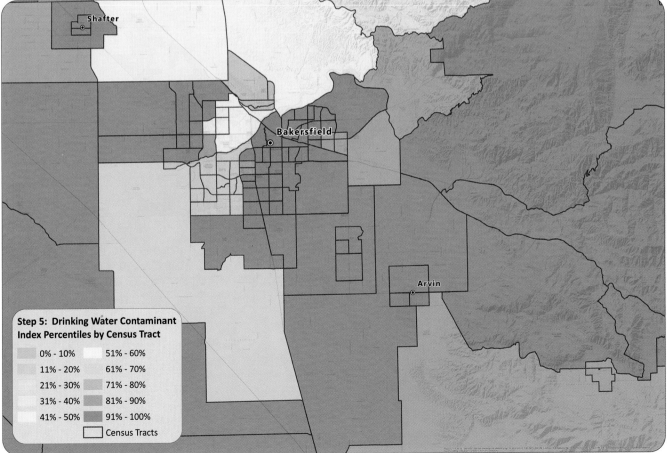

Step 5: Drinking Water Contaminant Index Percentiles by Census Tract

0% - 10%	51% - 60%
11% - 20%	61% - 70%
21% - 30%	71% - 80%
31% - 40%	81% - 90%
41% - 50%	91% - 100%

- Census Tracts

Brookings Water Distribution

City of Brookings

Brookings, Oregon, USA
By Jordan Fanning

Contact
Jordan Fanning
oregonjordan@gmail.com

Software
ArcGIS 10.2 for Desktop

Data Sources
CIty of Brookings, Curry County, Oregon Department of Geology and Mineral Industries, US Geological Survey

The City of Brookings is a small coastal community in Oregon with a resident population of about 6,500 and a utility customer base of about 3,300 homes. This water distribution map was created to be included with the city's quinquennial water master plan. It is the first official GIS- based utility map for Brookings and represents a benchmark for the municipality's commitment to technology for greater efficiency and information management. The map displays pipeline feature data, and shading tank and reservoir service areas on a parcel basis. This information provides utility workers with an at-a-glance understanding of how to micromanage network flow. This map accompanies a larger collection of utility field atlases and infrastructure maps. The City of Brookings has created a single database for accruing and maintaining all of its critical utility data. Data for the map was collected using many remote-sensing mediums, including lidar, mobile mapping, orthophotos, and hand-held GPS.

Courtesy of City of Brookings.

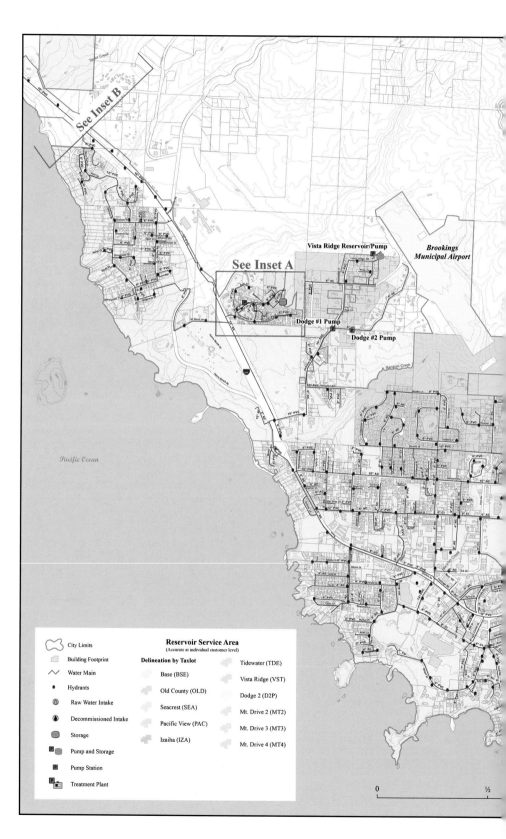

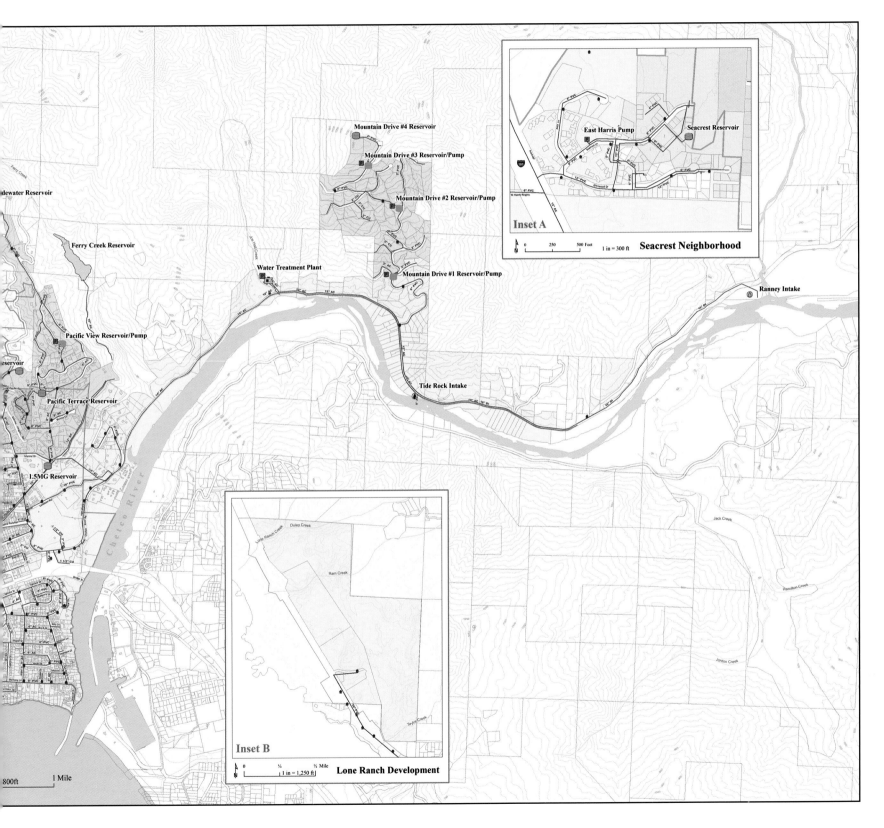

Mountain Drive #4 Reservoir

Mountain Drive #3 Reservoir/Pump

Mountain Drive #2 Reservoir/Pump

Mountain Drive #1 Reservoir/Pump

Water Treatment Plant

Ferry Creek Reservoir

dewater Reservoir

Pacific View Reservoir/Pump

eservoir

Pacific Terrace Reservoir

1.5MG Reservoir

Chetco River

Tide Rock Intake

Ranney Intake

Jack Creek

Hamilton Creek

Junton Creek

Inset A

East Harris Pump

Seacrest Reservoir

N

0   250   500 Feet

1 in = 300 ft

Seacrest Neighborhood

Inset B

Duley Creek

Lone Ranch Creek

Ram Creek

Taylor Creek

N

0   ¼   ½ Mile

1 in = 1,250 ft

Lone Ranch Development

800ft   1 Mile

Water Main Replacement Areas

Murray City

Murray, Utah, USA
By Matt McQuiston

Contact
Matt McQuiston
mmcquiston@murray.utah.gov

Software
ArcGIS 10.2.1 for Desktop

Data Source
Murray City Water Department

Murray City, Utah, located south of Salt Lake City, has adopted GIS technology throughout city departments. Murray City Water Department personnel have tracked main line breaks along with pipe size and material since 1999. This map shows hot-spot areas for main line breakage and hot-spot areas that have been remedied since 1999. The department serves 35,000 customers with 185 miles of water lines.

Courtesy of Matt McQuiston, Murray City

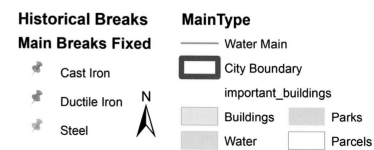

Historical Breaks

Main Breaks Fixed

Cast Iron

Ductile Iron

Steel

MainType

Water Main

City Boundary

important_buildings

Buildings   Parks

Water   Parcels

INDEX BY ORGANIZATION